NAVIGATING REGIONAL DYNAMICS IN THE POST-COLD WAR WORLD

I would like to dedicate this dissertation to my parents for their enduring drive and relentless love and support.

I currently identify myself as a Mediterranean Regional and International Relations Specialist. I am Lecturer at the Mediterranean Academy of Diplomatic Studies, University of Malta. I am also International Representative of the International Office, University of Warwick, England, and Book Reviews Editor of the journal Mediterranean Politics published by Frank Cass, London. I am also a member of several International Relations organizations including the International Institute of Strategic Studies (IISS), in London and the International Studies Association (ISA) in the United States. Author of several articles on Mediterranean and International Affairs in refereed publications and a regular news analyst in the international syndicated press.

Navigating Regional Dynamics in the Post-Cold War World

Patterns of Relations in the Mediterranean Area

STEPHEN C. CALLEYA
Mediterranean Academy of Diplomatic Studies, Malta

Dartmouth

Aldershot • Brookfield USA • Singapore • Sydney

I hereby declare that all the results presented within this dissertation were obtained by myself under the supervision of Professor Barry Buzan, with the exception of those instances where the contribution of others has been acknowledged. These results have not been submitted for a degree at any other institution.

Published by
Dartmouth Publishing Company Limited
Gower House
Croft Road
Aldershot
Hants GU11 3HR
England

Dartmouth Publishing Company
Old Post Road
Brookfield
Vermont 05036
USA

British Library Cataloguing in Publication Data
Calleya, Stephen C.
   Navigating regional dynamics in the post-Cold War world :
   patterns of relations in the Mediterranean area
   1.Regionalism - Mediterranean Region 2.Mediterranean Region
   - Politics and government - 1945-
   I.Title
   320.5'49'091822

Library of Congress Cataloging-in-Publication Data
Calleya, Stephen C.
   Navigating regional dynamics in the post-cold war world : patterns
   of relations in the Mediterranean area / Stephen C. Calleya.
      p.   cm.
   Includes bibliographical references.
   ISBN 1-85521-850-X (hb)
   1. Mediterranean Region--Foreign relations--1945- 2. World
   politics--1989- 3. Regionalism--Mediterranean Region. I. Title.
   DE100.C35 1996
   909'.09822--dc20   96-43126
                                                       CIP

ISBN 1 85521 850 X

Printed in Great Britain by Ipswich Book Co. Ltd., Ipswich, Suffolk.

Table of Contents

List of Tables, Graphs and Figures

Tables

Graphs

Figures

Foreword by Richard Falk

Despite the strong recent academic and journalistic interest in the prospects for further European integration, the dominant focus of inquiry in the discipline of international relations has continued to be on the sovereign state as primary actor and the states system as the foundation of world order. To the extent this statist outlook is challenged at all in the mainstream literature, it has tended to be contrasted with one or another scheme for world government, and if not in a full-blown utopian format, then in the shape of a significant strengthening of the United Nations. In essence, theorizing about international political life, and even policy advocacy, has been preoccupied for decades with the simple interaction between the part(state) and the whole(the world), with little systematic attention to intermediate possibilities.

Such a dichotomous conception of international relations is under challenge in a number of respects, first of all as a result of empirical tendencies. The state seems increasingly enfeebled by the dynamics of economic globalization and a variety of transnational pressures including those associated with social movements, drugs, crime, terrorism, pollution, migration, media transmission, computer networking. As a consequence, the territorially bounded sovereign state, although undeniably still formally in control of most international arenas and possessed, as ever, of awesome capacities to mobilize resources and engage in destructive warfare, no longer can credibly claim to be the only significant actor on the global stage. Furthermore, the challenges confronting most societies since 1989 are less and less centred on the residual strengths of the state as military powerhouse or holder of exclusive ceremonial status in most diplomatic arenas. In addition, especially in the aftermath of the Cold War, the United Nations, despite a deceptive flourish during the Gulf War, has resumed its position at the margins of international politics. Such resumed marginality was especially disappointing for idealistic champions of globalism who had discounted earlier limitations on the United Nations as mainly reflecting the gridlock produced by the Cold War and the pervasive rivalry of the two superpowers. To some extent, then, neither the statist, nor the globalist, levels of analysis now seems to suffice as a basis for responsible inquiry into the character of international political life. Both statism and globalism remain necessary, of course, but more is needed to take proper account of an increasingly complex, fragile, and multi-dimensional international reality.

If the focus is conceptual, then the two most significant enlargements of the field of inquiry are transnational social forces and regional actors, both of which in their distinct ways are expressive of the decline of statism. Transnational social forces, as in relation to environmental and human rights issues, are often a direct challenge to state power as exercised through the agency of governmental institutions, but often, also, a congenial supplement as in a variety of humanitarian emergencies where such informal external actors are welcome and can be effective, without encountering the sort of resistance that states or international institutions would meet.

Regional initiatives and regional actors are more closely and ambiguously linked to the state. With respect to the economic and social agenda of societies, regionalism is mainly, if not exclusively, an instrument of the state, often more so than for the citizenry itself. In Europe, for instance, the opposition to the ratification of the Maastricht Treaty, starting with the shock of the initial Danish repudiation of Maastricht in a referendum vote of June 1991, generally pitted the nationalist inclinations of society against the regionalist logic of the elites who were more receptive to the functional advantages of shifting power and authority to Brussels, even though it was largely *their* power. I recall a visit to Princeton by the Austrian Foreign Minister in this period who was assuring his audience that "Austria had always favoured the European Union, and would support Maastricht". When I asked, "Well, what of the Austrian people?" his answer was revealing: "Yes, it is true, 70% of Austrians are opposed!" Besides raising the question "Who is Austria?" such an exchange illustrates the close identification that has emerged in many instances between the promotion of regionalism and the viability in the late twentieth century of statism as the basic organizing principle of world order. In essence, there are two intertwined ways to interpret regionalism: as an extension of statist capacities to cope with global integration and interdependencies; as a superseding of statism, primarily as an intermediate defensive reaction against the threat of economic globalization. Both perspectives are aspects of the role of regionalism as an ingredient of the adaptive capacities of the embedded world order system.

There is a further matter that bears on the study of regionalism. It is the anxiety felt by elites of countries situated at the margins of regional formations. Such a phenomenon is best observed in Europe where regionalism, although currently on hold, has advanced the most, and is promising much more, especially in relation to money and banking. This anxiety can take two basic forms. It can involve the fears of submersion that are associated with full regional participation, and it can involve the fears of decline and functional disadvantage that are associated with exclusion and non-participation.

It is against this intellectual and policy background that one appreciates this immensely valuable study of Stephen Calleya. It contributes in fascinating ways to our understanding of the general issues raised by the emergence of regionalism as a policy alternative for states situated at the margins of Europe, and explores these generalities by a critical analysis of the prospects for converting the Mediterranean into an international region. Such a concrete exploration is not a figment of a fevered scholarly imagination, but represents a response to lots of loose talk and vague undertakings that have projected the image of the Mediterranean as a viable region, which if so structured, could bring peace and prosperity to this troubled area of the world.

Calleya explores these concerns with impressive mastery of the academic landscape, including its various theoretical centres of controversy, but also in the spirit of scholarly detachment. He reaches sceptical conclusions about the regional project in the Mediterranean, but he does so after a careful assessment of the positive case, including an examination of the glorious Mediterranean past when its psychological and material reality did constitute a regionalist construction of political life, above all during the long reign of *Pax Romana*. Among the many conceptual contributions made by Calleya, his emphasis on component sub-regions may be the most impressive and distinctive. Calleya shows persuasively that the Mediterranean as a region is most usefully conceived of as consisting of three sub-regions: southern Europe, Levant (Eastern Mediterranean), and Maghreb (Southern Mediterranean). What he shows is that these sub-regions do not have convergent outlooks or priorities that would make any type of serious integration a political option in the foreseeable future. To the extent that regionalism is attractive at all, and it is for many of them, it is to be included in the non-Mediterranean regional system evolved by the countries of Western Europe. In this sense, the advocacy of a distinct Mediterranean regionalism seems merely like a sentimental exercise of politicians dreaming of recreating the past or finding the pretext for yet another round of meetings that eventuate in the establishment of a new haven for bureaucrats in the form of a secretariat. Even Malta, the arch advocate of the Mediterranean idea, is far more concerned with being accepted as a member of the European Union than it is with the chimera of an alternative regionalism that links southern Europe, the Levant, and the Maghreb.

Calleya goes further than providing a critique. He explores the reality of the Mediterranean circumstance, and concludes convincingly that its geopolitical destiny is to serve as "a frontier" rather than as a "region". By this he means specifically the divisions that exist with respect to civilizational identity (Christian/Islamic), stage of development, political orientation(constitutional democracies/authoritarian governments), and geography (Europe/Africa; Europe/Middle East or West Asia). The Mediterranean is a crossroads, not a new regional locus for the deepening of cooperative relations among participating countries. This reorientation of Mediterranean identity has enormous implications, including providing diplomats in the area with a creative outlook on their situation that can displace the dead end pieties of an entirely fanciful Mediterranean regionalism.

Stephen Calleya has given us far more than a critical case study of Mediterranean regional pretensions, although he has done this with consummate skill. The wider value of his book is to provide a model for the assessment of regional prospects under varying geographic constellations that exist today. Given the pressures on states arising from globalization and transnationalism, the regional option is obviously a tempting adaptive strategy. What Calleya helps us understand is that regionalism, no more than globalism, is a panacea for the ills of the states system. It can, to be sure, enhance the participation of particular states and groups of states in relation to prevailing patterns of the world political economy, but not invariably or necessarily. The positive case for regionalism must be made in relation to such contextual factors as shared history, civilizational identity, level of economic development, foreign trade and investment patterns, degree of conflictual relations, quality of leadership, public opinion, intrusion of geopolitical influences. If these factors do not encourage *specific* regional initiatives, then their affirmation will only produce disappointment and add to the general mood of disrespect for the capacities and integrity of public officials.

In the end, then, Stephen Calleya has with this book given us an all-purpose framework for a sophisticated comprehension of the regional dimension of world order. His brief is not one of advocacy, but rather one of understanding.

Richard Falk
Albert G. Milbank Professor
of International Law and Practice
Centre of International Studies
Princeton University

June 1996

Preface

By providing a holistic viewpoint this book has gone some way to correct the imbalance in the secondary literature which has been dominated by coverage of the Mediterranean from either a domestic or international perspective.

The study clarifies whether there has been a re-surgence of regionalism in the Mediterranean since the end of the Cold War by examining regional dynamics in the area from five different perspectives. The first investigates the concept of regionalism in international relations. The second gives an historical perspective of regionalism in the Mediterranean. The third focuses on contemporary international relations in the Mediterranean. The fourth examines the influence of external actors in the international politics of the Mediterranean. The fifth provides a "reality check" of the Mediterranean in the post-Cold War world.

Patterns of interaction in the Mediterranean area reveal that there are two neighbouring international regions that are very different in character: Western Europe and the Middle East, which include three subregions bordering the Mediterranean, Southern Europe, the Levant, and the Maghreb. Post-Cold War trans-Mediterranean political proposals resonate with older traditions. This research project investigates the substance behind such Mediterranean regional rhetoric. The conclusions of this analysis are that external actors have often influenced regional patterns of relations, but have not been able to alter the basic pattern of regional alignment and conflict within international regions. In addition, regionalizing proposals are more accurately described as attempts to pursue particular national and subregional interests, and as boundary management devices rather than boundary transcending ones. A reality check at the end of the twentieth century shows that the Mediterranean is more of a frontier zone than an international region.

Throughout the three years of research on this project I have focused on secondary literature relating to the concept of regionalism in the discipline of international relations under the guidance of my supervisor, Professor Barry Buzan. I also reviewed primary and secondary documents referring to security initiatives in the Mediterranean. I achieved this by conducting specialized study and consultations at several British and international academic institutions which include the International Institute of Strategic Studies, the School of Oriental and African Studies (SOAS), the Royal Institute of International Affairs at Chatham House and the British Library in London, St. Antony's Library and the Bodleian Library in Oxford, Warwick University Library, the Hellenic Foundation for European and Foreign Policy in Athens, the Mediterranean Academy of Diplomatic Studies in Malta, the NATO Library and Policy Planning Section in Brussels, the Institute for Security Studies of the Western European Union in Paris, the Centre for Peace and Disarmament at the Universitat de Barcelona, the Council on Foreign Relations in New York, and the Library of Congress in Washington D.C.

A significant body of material concerning contemporary international relations in the Mediterranean was obtained by participating at numerous international conferences contending with security challenges in the Mediterranean. These include a Hellenic Foundation Conference in Halki, Greece in the summer of 1993, an ECPR Conference in Madrid in April 1994, a Conference at Reading University in November 1994, a European Peace Research Association (EUPRA) Conference in Malta in November 1994, the International Studies Association Annual Convention in Chicago in February 1995, and two Wilton Park Conferences in February and April 1995. I also conducted several interviews with officials involved in policy-planning initiatives in the Mediterranean which include, David Law, former Head of the Mediterranean Unit at the NATO headquarters in Brussels, Guido De Marco, Maltese Minister of Foreign Affairs, John Roper, Director of the WEU Institute in Paris, Roberto Aliboni, former Director of the Instituto Affari Internazionale in Rome, Theodore Couloumbis, Director of the Hellenic Foundation for European and Foreign Policy in Athens, and Nick Carter, former Head of the Southern European Desk at the Foreign and Commonwealth Office in London.

Acknowledgements

I am grateful to the Foreign and Commonwealth Office in London and the Rotary Club International for helping me meet the initial financial costs this endeavour has entailed. I also thank the academic staff in the Department of Politics and International Studies at the University of Warwick for their assistance and support, in particular, Barbara Allen Roberson, Charles Jones, Wyn Grant, Susan Strange, and Peter Burnham. I also thank Richard Gillespie and Claire Spencer for their support. Special thanks also go to the staff at the International Office at the University of Warwick, in particular the Director, Antony Gribbon, and the staff at Warwick University Library and Computer Services for their continuous assistance. I also thank all the staff at the Mediterranean Academy of Diplomatic Studies, University of Malta, for their support. I am also grateful to my brother Peter and my colleagues Richard Stables and Annabel Kiernan for their assistance and encouragement throughout my research project at Warwick. Thanks also to Professor Richard Falk who kindly agreed to write the Foreword to this book.

Most of all, however, a special thanks to my supervisor Professor Barry Buzan, for bringing his corrective expertise to bear on a much scribbled research draft. The final product has been improved a great deal as a result of his endless advice and support.

List of Abbreviations

5 + 5 - West Mediterranean Forum

AFSOUTH - Allied Forces of Southern Europe

AMU - Arab Maghreb Union

CIS - Commonwealth of Independent States

CM - Council of the Mediterranean

CSCE - Conference on Security and Cooperation in Europe

CSCM - Conference on Security and Cooperation in the Mediterranean

EC - European Community

EEA - European Economic Area

EFTA - European Free Trade Area

EIB - European Investment Bank

EU - European Union

FDI - Foreign Direct Investment

FIS - Front Islamique du Salut

GATT - General Agreement on Tariffs and Trade

GCC - Gulf Cooperation Council

IGC - Intergovernmental Conference

IMF - International Monetary Fund

NACC - North Atlantic Cooperation Council

NAFTA - North American Free Trade Area

NATO - North Atlantic Treaty Organization

OSCE - Organisation on Security and Cooperation in Europe

PFP - Partnership for Peace

PHARE - Poland Hungary Aid for Reconstruction (of the EC)

PLO - Palestine Liberation Organisation

SADR - Saharan Democratic Republic

UN - United Nations

UNCED - United Nations Conference on Environment and Development

WEU - Western European Union

1 The Study of International Regions in the Post-Cold War World

Introduction

This book deals with the issue of international regions in the aftermath of the Cold War. The study of the international politics of regions has often been described as being underdeveloped (Buzan, 1991a: 198; Neumann, 1994: 53). This thesis seeks to clarify the confusion surrounding the theories, models, paradigms, and analytical frameworks already existing in this field. The aim of this introductory chapter is to summarize the main trends and approaches in the study of regionalism. It will also be demonstrated that the 'regionalizing world' theme has re-emerged as a dominant topic in contemporary international relations literature and a brief survey of recent references to regionalism in the Mediterranean area will be conducted.

1.1 Main Trends and Approaches in the Study of Regionalism

Although the parameters of the post-Cold War world are still in a state of flux, it seems apparent that the bipolar international system of the past fifty years is being replaced by a more multipolar configuration. The demise of the Soviet Union and the limited withdrawal of the United States from the centre stage of international affairs has ushered in a period where international power relations are more erratic. The equilibrium which the bipolar Cold War provided, through its configuration of blocs, alliances, stalemates and sustained crisis, has been superseded by a fluid system. A distinct feature of the present transformation is that regional politics have gradually gained in prominence.

As the post-Cold War label suggests, the 1990s are not a period completely different from that of the Cold War days. It is characterized by a

1

level of continuity and discontinuity with the previous bipolar international system. Contemporary international relations tend to be neither moving in the optimistic direction of a new world order nor in the pessimistic one of "chaos" and "turbulence" as has been suggested (Rosenau, 1992). Instead, the dynamics of international relations appear to be going through a transitory phase, which one author describes as "disorder restored" (Mearsheimer, 1992: 213-37). Cantori recently described the post-Cold War pattern of relations as a dynamic multipolar international system based on regions and regional subordinate systems within which the struggles of nationalist identities and hegemonic leadership rivalries are taking place (Cantori, 1994: 22). From a regional perspective, this is illustrated by the fact that the principal change in the geopolitics of the world-system in the 1970s and 1980s has been the decline in the relative power of the United States. This trend is a key feature of the post-Cold War international system and is a normal cyclical occurrence (Wallerstein, 1993: 4).

The economic strengths of the European Union (EU) and Japan have been steadily increasing since the mid-1960s, and the United States has not been able to keep pace. As a result, U.S. foreign policy in the past two decades has been centred around ways to slow the pace of this loss of hegemony by exerting pressure on its allies to share the burden of global security.

The emergence of a more polycentric power structure at the system level is therefore certain to have a profound impact on regional relations. The superpower grip on international relations in some instances resulted in overlay. This occurs when the direct pressure of outside powers in a region is strong enough to suppress the normal operation of security dynamics among the local states (Buzan, 1991a: 198).

On some occasions overlay stifled interaction among regional actors by limiting the parameters within which they could operate. The United States' foreign policy in Central America and the Soviet's policy in Eastern Europe are examples of this development. Bipolar intervention quelled regional hostilities when it took the form of overlay, as in Europe. But when it took the form of direct intervention, as was the case throughout the Third World, it often amplified crises across a larger geographic area rather than subdued them (ibid.: 208).

The collapse of superpower overlay has led some international relations scholars to forecast that regional international politics will again become a dominant characteristic of the international system (Rostow, 1990: 3-7, see also Falk, 1995: 1-16). The World Bank has recently made the argument that economically, "regionalism is back and here to stay" (Melo and Panagariya,

1992: 1). The increase in regional agreements that have been reported to the GATT in the last five years tends to verify this trend (see Graph 1). In his economic analysis of the world's future, Kennedy proceeds by regional analysis and comparisons and by identifying likely regional winners (Kennedy, 1993). These analysts base their assumptions on the premise that a multipolar system of states, where power is more evenly distributed among a number of regional powers, is in itself conducive to the rise of regional politics (Buzan, 1991a: 207). Buzan illustrates this development by identifying that regional dynamics are evolving against a background in which the higher-level complex is also entering a period of transformation. In other words, the intrusive patterns from higher to lower levels which were characteristic of the bipolar rivalry between the superpowers (the creation of alliances, sustained crisis and military conflicts) have shifted as a result of the Cold War's breakdown.

Graph 1

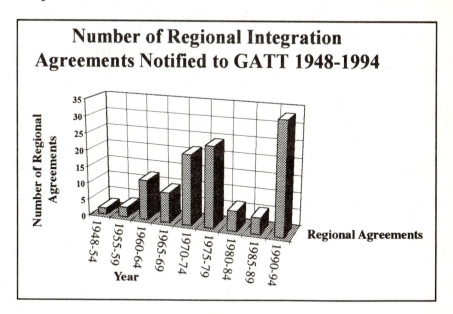

Source: *The Financial Times,* 27 April 1995: p.7

A plethora of paradigms have been conjured up by international relations scholars, to help explain the implications of the sea-change in international affairs since 1989. Huntington predicts that the end of the Cold War does not mark the end of history, or a return of traditional rivalries between nation-states, but a new phase in world politics where conflicts of global politics will occur between nations and groups of different civilizations (Huntington, 1993: 22). Nye, Jr. describes world order after the Cold War as *sui generis* which cannot be understood by "trying to force it into the procrustean bed of traditional metaphors with their mechanical polarities" (Nye, Jr. 1991: 88). Power is constantly becoming more multidimensional. Nye, Jr. states that the distribution of power in world politics has become like a layer cake. The top military layer is mainly unipolar, with the U.S. dominant in military power. The economic middle layer is tripolar and has been for two decades, with the United States, the European Union, and Japan governing. The bottom layer of transnational interdependence shows a diffusion of power to the peripheral sector of the international system (ibid., and Nye, 1991: 191-2).

Buzan agrees that the diffusion of power in the contemporary system raises the importance of actors at the bottom of the power hierarchy for the functioning of the system as a whole. A reduction in the intensity of global political concerns and a decrease in the resources available for sustained intervention is conducive to the rise of regional politics (Buzan, 1991b: 435). Buzan adopts a structural realist approach and a centre-periphery model of the international system to depict the changes taking place. In his model of concentric circles, the "centre" implies a globally dominant core of capitalist states; the "semi-periphery" consists of the more robust and industrialized states in the periphery; while the "periphery" is a set of politically and economically weaker states operating within a network system mainly created by the centre (ibid.: 432).

A review of the various models noted above, reveals that there is no consensus about the fact that diffusion of power from the centre to the periphery is taking place. Although the pattern outlined above strongly suggests that a different pattern of geopolitics is evolving in the world, there is not enough consensus among international relations scholars on whether the advent of a multipolar system will increase or decrease the likelihood of regional conflicts.

Cantori and Spiegal suggest that a multipolar structure is the system most conducive to amplification of local conflicts. In areas where the superpowers had a strong degree of influence over the foreign policy agenda of client states, like U.S. dominance over South Korea, or the Soviet Union over Cuba, a high degree of co-operation took place. When either the United States or the Soviet Union lost power in international regions, hostilities increased and even led to outright conflict in places such as the Middle East, North Africa, South-East Asia, and the Gulf (Cantori and Spiegal, 1970: 33-7).

Buzan counters that although a multipolar architecture permits for a more uncertain chain of events, it is also less intense than a bipolar zero-sum competition. In addition, a multipolar matrix grants regional powers more room to manoeuvre when conducting their external political relations. Hence, the disappearance of superpower rivalry and the decentralization of power throughout the international system are two tendencies that forebode an era where great power intervention in regional affairs is likely to diminish (Buzan, 1991a: 208).

Another factor that supports the regional resurgence school of thought is that for the first time the seeds sown during the period of decolonization are in a position to develop now that the superpower strait-jacket has been removed. In his study on regional conflict formations, Vayrynen cites the 1950s as the starting point of the rise of regionalism time-line (Vayrynen, 1984: 339 and 345). At the time, integration efforts in the industrialized world were mainly sponsored by the great powers.

Decolonization, the second main phase, paved the way for early efforts at regional economic integration and regional security arrangements in the Third World. Economic integration was mainly a response to the economic predominance by industrial powers. Security groupings tried to cope with the departure of colonial powers and the multiplicity of newly independent nation-states and their military establishments. The most recent phase of regionalism is a manifestation of the effective spread of power from the core of the international system to its peripheries. It has given rise to new regional power centres, integration schemes, and regional conflict formations.

Whether the absence of the superpower constraint will mute or exacerbate regional conflicts will very much depend on the circumstances of each particular regional grouping. In some areas it may increase the intensity of conflict as rivals attempt to settle long subdued scores, as is the case in the former Yugoslavia. In others, as the example of the Middle East demonstrates, it may encourage external assistance for regional endeavours at conflict resolution. In any case, post-Cold War developments support the theory that the intrinsic dynamics of subordinate systems are becoming more significant features of the international system, a trend that is likely to continue as technological advancements are more widely dispersed (Buzan, 1991a: 208).

The recent increase in references to the 'regionalising world' theme by scholars of international relations is reflective of the multiplicity of regional agreements that have been signed since the end of the Cold War (see Graph 1).

China has already established itself as an independent great power in Asia, with India and perhaps South Africa poised to become dominant actors in their areas of the world. The gradual retrenchment of the United States and Russia in Europe and Northeast Asia, presents both the European Union and Japan with the opportunity to play roles in the international system commensurate with their power (ibid.: 207-8). Despite this development, can we talk about the resurgence of regionalism in international relations? Or are we perhaps confronted rather by derivations and extensions of constellations and potential conflict, the nuclei of which lie elsewhere? What theoretical explanation do we have to interpret this unit of analysis? In addition, what are the empirical indicators of regional transformation?

In an attempt to answer these questions this research project provides a schema for the analysis patterns of interaction within and between international regions. This will include defining the term "international region" and establishing ideal types of regionalism which will assist in the task of identifying tendencies towards regionalism. In this sense, the thesis adopts the framework of regional system theorists who attempt to understand what role such factors as symmetry or asymmetry in capacity and similarity or dissonance in ideology, play in encouraging conflict or co-operation in regional international politics. When international regions interact, particular regularities occur, and under certain specific conditions certain cause-effect relationships become more apparent. The empirical systems approach encourages the postulation of particular types of hypotheses for investigation. Each question put forward will assist in identifying which factors enhance or constrain the evolution and development of international regions.

For example, do certain international regions stimulate, induce, or contribute to the formation of other international region types? This will include examining which types of foreign policy interaction are more likely to nurture similar models of international regions. An international region can influence other states in a number of ways: it can be perceived as a threat against their self interests (the "threat effect"), or an international region can be perceived as a role model to other group of states (the "demonstration effect").

In the first instance, an international region assists in the creation of other regional types whenever other states feel that their interests are being undermined. One could for example argue that the North American Free

Trade Area is mainly an initiative aimed at countering the integration efforts being pursued by the European Union member states. In the second instance, an international region may simply be imitated because it has served as an adequate model for problem-solving or it has demonstrated how a group of states can enhance their international weighting by participating in a larger network system (the demonstration effect). This may help explain the emergence of numerous regional integration schemes over the years that in some manner or other have tried to replicate the process of European integration that began in 1957 (Helleiner, 1994: 1-5).

These questions expand our field of concern from the task of international region formation to that of interaction after such systems are functioning. An initial assessment reveals that there are no specific patterns observable in this field of research - for instance, at an economic level, how do certain regions react to the elaboration of specific policies of multinational companies vis-à-vis those regions? Does economic interaction between societies in an international region stimulate responses on a regional level somewhere else? The interaction between transnational dominant international regions at first appears important as this is an area of social action in which one of the constituents of "interdependence" is located. Both economic theory, such as Viner's concepts of "trade creation" and "trade diversion", and communications theory (employed to measure the intensity of relations in trade or in shared membership in international organizations) have registered some of the phenomena in transnational society, but have not been able to explain their social and political relevance, let alone their causes.

Focusing attention on the interaction between already functioning international regions reveals that no single recurrent pattern of interaction is discernible. In general, two patterns of interaction appear to occur frequently between intergovernmental dominant types of international regions in the field of "high politics" (high politics is here referring to matters of national prestige, survival, and defence or power status in international politics; the main international regions types are elaborated on in the framework of analysis in section 2.3, but it is not entirely clear what triggers these patterns to occur or why they sometimes appear to function simultaneously).

The first type of interaction is the symmetry pattern: international regions interact in such a way as to produce similar processes within both systems. The second type of interaction is the contrast pattern: international regions interact in such a way as to produce different, if not opposite, processes within each region.

The symmetry pattern is observable under the following conditions:

international regions tend to reinforce each other if i) there exists a mutual involvement between the societies of the two international regions, ii) regional processes within one system hurt the interests of the other, and iii) there is no real possibility of participation in the other international region. Once again the "threat effect" constitutes the main transmission belt. In less abstract terms, this hypothesis suggests that the success of one international region is detrimental to the development of another similar system. This effect forces the latter to co-ordinate policies among states in the area and to arrive at a consensus on how to maximize their comparative advantages. In his analysis of the Central American Common Market, Schmitter suggests that relations with the international community are the primary factor that influenced the development of integration once a certain level of internal integration had already been achieved in what he defines as an "externalization hypothesis" (Kaiser, 1968: 100).

The functioning of this pattern can also be observed in the interaction between the European Union and other industrialized countries (primarily the United States). Throughout the GATT Uruguay Round, and especially in the months leading up to the set deadline for negotiations, pronouncements often reached quasi-hostility and the Round often looked like a regulated conflict between international regions.

The contrast pattern is observable in the following circumstances: international regions weaken other similar systems if i) they hurt the interests of members of other international regions, ii) they are reasonably strong and offer rewards for participation, and iii) participation in or access to the benefits of the international region is an option that is reasonably open to individual members of the other international regions.

In the Mediterranean area, one can argue that the European Union has weakened the prospects of other regional movements in the vicinity such as the European Free Trade Area and the Arab Maghreb Union. The opportunity of direct association to a certain number of European and Arab states to participate in the benefits of this comprehensive international region is enough to discourage the states concerned from pursuing the establishment of their own international region. Hence, by offering sufficient enough rewards and remaining "open" enough to make participation through membership, association, or special arrangement a reasonable alternative, the European Union has hindered other regional efforts. (The impact of the European Union on the Mediterranean is discussed in detail in section 4.4.)

States which are confronted with the dilemma of either joining such an "open" international region or attempting to form their own international region have two alternatives to choose from: they can either strengthen their own grouping (the contrast model), or they can individually seek direct accommodation with, or participation in, the other international region (the symmetry model). If they opt for the second alternative, however, they do so at the cost of undermining their own international region, and, potentially upsetting bilateral relations.

Contrary to the criticism of a "Fortress Europe", which stresses the threat of adopting an inward-looking or closed model, an international region can become a threat to the survival of other regional groupings when it is prepared, or potentially willing, to concede participation in the benefits of its system to "outsiders". Such a possibility presupposes that there are real rewards for participation, and that it cannot take place if there is a significant geographical distance between the two areas coupled with an absence of closer political ties. Hence, while Latin American countries have had little choice but to pursue the contrast pattern in order to counter the EU integration process, a number of African countries, tied to the EU by bonds dating back to the colonial period, are seeking the symmetry course through participation in association agreements.

Each hypothesis assists in assessing the specific features of any particular international region and when applied to the Mediterranean area, will facilitate in the complex task of identifying regional demarcation lines. The complexities involved in formulating such a basis of analysis is one of the reasons that has discouraged research into this area over the last two decades. In particular, power theorists have tended to underplay the significance of the regional level in international relations (Buzan, 1991a: 186). During the 1970s and 1980s attention focused on relations between the superpower dyad, with regional dynamics largely discussed as an extension of this power struggle. However, the challenge which one must confront when attempting to explain patterns of regional interaction and regional transformation should be regarded as part of the process of developing a coherent theory that attempts to provide a framework categorizing and illustrating the dynamics of such a complex research question. This problem has been increasingly prioritized in the post-Cold War era. The collapse of superpower influence in international relations, coupled with the rise of independent regional actors, implies a new epoch in which the region will again become one of the crucial traits of international politics.

1.1.1 The Case of Regionalism in the Mediterranean Area

This scheme of analysis has been applied to the Mediterranean basin in order to examine the dynamics of regional international politics in the post-Cold War world. As a conceptual entity, the Mediterranean area eludes a coherent and comprehensive definition. Contemporary political and academic interpretations are susceptible to subjective reasoning. For example the European Union defines the Mediterranean as countries with a Mediterranean coastline plus Jordan. The inclusion of the latter transforms the logic of a geographically based definition into one affected by political considerations (Henson, 1994: 1). Much like Europe, the Mediterranean suffers from the vagaries of geography: just how far north, south, east, and west should one take the boundaries of the Mediterranean area? Geographically, the concept of a Mediterranean area is primarily limited to the three regional subgroupings which encompass the basin, namely Southern Europe, the Maghreb and the Levant. However, throughout this study, the concept of a Mediterranean area is geopolitical in nature. It represents all countries which border the Mediterranean sea plus those countries with an interest in the area. The regional dynamics operating in Central and Eastern Europe and especially the Black Sea area also impact on the littoral states of the Mediterranean but are excluded from this study for two reasons. First the patterns of interaction between these subregions and the riparian states of the Mediterranean are of a largely limited nature. Second, inclusion of this group of countries would only add to the already elaborate schematizing framework that has been set up without contributing any different dimension to the overall investigation. Thus attention is specifically focused on the pattern of relations within and between the Western European and Middle Eastern international regions and external actors with an interest in the area such as the United States.

The main reason the Mediterranean area was selected as a case study is due to the fact that in recent years there has been a rapid increase in Mediterranean regional rhetoric. The purpose of this study is to investigate what substance, if any, is behind this rhetoric.

Given that the Mediterranean is an area whose varied regional parameters cannot be easily reconciled in a single equation, the framework of analysis proposed seeks to identify the tendencies that contribute to regional variation in the Mediterranean. This is sought by applying the threefold scheme of analysis, described at the end of chapter two, to the various phases of Mediterranean history, from Antiquity through to contemporary times.

Since the collapse of the Cold War a series of conflict prevention and peace-building proposals have been presented by Mediterranean countries in an attempt to promote the notion of a 'Mediterranean region'. The breakthrough in September 1993 in the Middle East peace process somewhat accentuated this effort. Unless all interested parties in the littoral are involved in region-building deliberations, the goal of establishing a trans-Mediterranean regional network cannot be realized. A thorough analysis of these initiatives can be found in chapter five. An assessment of pan-Mediterranean proposals will require applying the framework of analysis to the Mediterranean area to detect which patterns of interaction are taking place. As a means of introduction, a brief synopsis of the numerous contemporary proposals follows.

The most far reaching trans-Mediterranean regional proposal put forward since 1989 is the Italian-Spanish project. The proposal was officially presented in September 1990 and envisages the involvement of all Mediterranean states (from Mauritania to Iran) in a Conference on Security and Cooperation in the Mediterranean (CSCM) (Italian-Spanish CSCM Non-paper, 1991; Ordonez, 1990: 1-8; Martinez, 1991: WEU Doc. 1271).

A yet more intensive sub-set of regional dialogue can be found in the French initiative, established in the western sector of the Mediterranean: the Western Mediterranean Forum. The Forum was officially launched at a foreign ministerial level in Rome on 10 December 1990 and was composed of four Southern European members (France, Italy, Portugal, Spain) and five North African members (Algeria, Libya, Mauritania, Morocco and Tunisia). The Forum developed into the "5+5 talks" when Malta joined the process in 1991 and attended the follow-up meeting in Algiers in October 1991. The third meeting that was scheduled to take place in Tunis in 1992 has not yet materialized. This is largely due to the breakdown in relations with Libya because of the Lockerbie affair and the deterioration of political stability in Algeria. For the time being, Libyan intransigence and the Algerian domestic crisis have relegated the French initiative to diplomatic limbo (Joffe, 1994: 29-30).

A succession of other efforts at creating a trans-Mediterranean regional network have been proposed. In June 1992 a CSCM-type meeting took place at Malaga and adopted a final document that was divided into three sections: a security chapter, a co-development and partnership chapter, and a human rights chapter. The Malaga document also incorporated a preamble stating that although the conference was not mandated to devise direct solutions to conflicts in the Mediterranean, the purpose was "to launch a pragmatic process of co-operation which will gradually gain in strength and coverage, generate a positive and irrepressible momentum, and facilitate the settlement of these conflicts" (Ghebali, 1993: 100-1).

Although the Conference on Security and Cooperation in Europe (CSCE) avoids mentioning the CSCM project in the Helsinki Declaration of July 1992 entitled The Challenges of Peace, it still recognizes the significance of the Mediterranean to the future of European security:

> We recognise that the changes which have taken place in Europe are relevant to the Mediterranean region and that, conversely, economic, social, political and security developments in the region have a direct bearing on Europe (CSCE Helsinki II Declaration, July 1992: Point 37).

Despite the indifference manifested by the CSCE (since December 1994 the Conference has become the Organisation for Cooperation and Security in Europe, OSCE) towards the Italian-Spanish CSCM project, it also organized a CSCM-type meeting in Malta in May 1993. In fact, the CSCE seminar was held in accordance with the provisions of the Helsinki Document of 1992 (CSCE Mediterranean Seminar's Chairman's Report, 1993). As was the case at Malaga, the Seminar's participants were free to discuss problems pertaining to the Mediterranean without the inhibitions normally associated with the onus of a negotiated text.

Throughout 1993, a group of Mediterranean countries led by Malta proposed the creation of a Council of the Mediterranean. The forum could be established upon the framework of the Council of Europe, creating facilities required to involve all the parties concerned in a continuous dialogue towards the solution of problems affecting the area. The Council would also incorporate the active participation of all integrated parties including the European Union, the Arab Maghreb Union, and the Arab League. As guidelines for membership, the following criteria have been drawn up: adherence to the principles of the UN Charter, respect for the dignity of the human person and the Rule of Law, and respect for the establishment and development of representative institutions (De Marco, 1993: 1-7).

When outlining the structure of the Council, proponents of the idea proposed appointing a Committee of Ministers, a general assembly with consultative powers where representatives of Mediterranean states can form a Parliamentary Assembly of the Mediterranean, supported by a secretariat intended to co-ordinate activities of the Council in the political, economic, social, environmental and cultural fields (ibid.). To date this initiative has only received lukewarm support from representatives of the Maghreb region, particularly Tunisia, for whom the establishment of a CSCM remains a priority. Yet the concept of a Mediterranean Council remains a viable option as it complements the new spheres of co-operation that the EU has highlighted in its Euro-Maghreb Partnership document, which is discussed at length in chapter four (Rato, 1994).

In November 1993 Italy proposed establishing a Mediterranean type "Eurocorps" (Atlantic News, 1993: 2). The corps would incorporate Spanish, French and Italian military and naval forces and would be the South's contribution to the pre-existing Eurocorps, which brings together French, German and Belgian military units. At the start of 1995, the three Southern European Union member states and Portugal established a joint land force called Eurofor, with a headquarters in Florence, and a joint naval force under French command, called Euromarfor. The new forces fall under WEU command and control structures but are also at the disposal of NATO (Financial Times, 2-6-95: 2).

The creation of such a security organization could exacerbate the socio-economic disparities that exist in the Mediterranean if it remained an exclusively European endeavour. On the other hand, it could assist in bridging the existing North-South divide in the basin if, for instance, it extended membership to non-European Union Mediterranean member states. But bridging entails the notion of two way traffic. The Southern Mediterranean states must be allowed to participate in the decision making process if co-operation in the area is to increase. NATO's North Atlantic Cooperation Council (NACC) and its Partnership For Peace proposal are models that a Mediterranean Eurocorps or "Medcorps" might consider imitating as a means of achieving such an objective (Financial Times, 27/28-12-93: 3). Both promote the active involvement of non-member states by granting observer status to interested countries in the organization's vicinity. One positive result is that there has been an increase in joint security ventures among the parties concerned which serves as a step towards a better understanding of the security concerns of the area.

In the post-Cold War climate, there are suggestions that region-building theories are re-emerging (Neumann, 1994: 53). After fifty years of stability under superpower constellations, secondary and more minor powers are having to reassess their strategies on the basis of the changes in the world power structure. In the seventies, interest in the field declined, mainly because difficulties arose in locating empirical proof of the theoretical hypotheses concerning both the dependent variable (region transformation) and the various independent variables (geographical size, political system, and the degree of development) in the equation. This study therefore attempts to go some way towards filling the research gap that has existed for the last twenty years in this field of study and put the issue of regionalism back at centre-stage. For this purpose, this research project is structured as follows.

The first and second chapters are conceptual and deal with three issues: various approaches to the study of the politics of international regions, a survey of the literature on international regions, and the framework of analysis to be applied in the case study. Chapter Three provides a historical perspective of regionalism in the Mediterranean. The aim of this chapter is to outline the way in which the Mediterranean has evolved as a region which will assist in evaluating on current regional tendencies.

Chapters Four and Five are concerned with current trends in the international politics of the Mediterranean. The former examines the internal dynamics of the international regional subgroupings bordering the Mediterranean area, namely, Southern Europe, the Maghreb and the Levant, by applying the framework of analysis to the three. The latter assesses future prospects for regional development in the Mediterranean. It assesses the various pan-Mediterranean initiatives that have emerged in the last few years in a particular effort to detect which tendencies, if any, point towards a resurgence of regionalism taking place in the Mediterranean in the aftermath of the Cold War.

Chapter Six assesses the influence of bilateral external intervention in the international politics of the Mediterranean. The penultimate chapter examines the relationship between international organizations and regional developments across the Mediterranean. Both chapters attempt to identify what impact the collapse of the Cold War is having on intrusive action in the Mediterranean.

The final chapter provides a "reality check" of the Mediterranean area in the post-Cold War era. It discusses the likelihood of the Mediterranean becoming an international region in the near future and the possibility of alternative patterns of interaction.

This structure then provides a guide to the complex issues involved in addressing the subject of regional interaction and transformation. It suggests the main themes for analysis and gives a comparative framework within which the research is conducted.

2 The Concept of Regionalism

2.1 A Literature Review of Regionalism

The aims of this chapter are to survey the literature on 'international regions', and to suggest a framework for analysis that takes into account the major conceptual contributions in international regions theory in the previous decades. The results of these theoretical investigations will later be applied to contemporary patterns of interaction in the Mediterranean in an effort to identify any regional tendencies in the area.

The notion of employing regions as a basis for international schemes to promote world order is a relatively recent exercise. Peace initiatives presented during the classical world order, of European-dominated international relations up until to the First World War, had very little or nothing to say about regionalism (Hinsley, 1963: 10). Official and academic thinking during the inter-war years continued to be universalistic rather than regionally based. It was only after the collapse of the League of Nations, and the return of alliance theory that regionalism became, in the 1940s, a major topic of discussion.

Regionalist thinking developed rapidly in the decades of the 1940s and 1950s. Prompted by the emergence of European regional organizations, regional research programmes flourished in the 1950s and 1960s. Interests in this area of study waned in the seventies as the European integration experiment slowed down. Another factor which contributed to the lack of academic interest in regionalism was the influence of the realist and later the neo-realist approaches to international relations. Throughout the 1970s and 1980s they dominated the study of international relations and stressed the importance of assessing interaction between the state and the international system. Very little or no regard was given to the region as a unit of analysis. The collapse of the Cold War structure and corresponding developments in the international political economy has renewed academic interests in the study of regionalism. This study aims to help achieve a better understanding of the process of international region formation and transformation in light of post-Cold War changes in the international system.

The fundamental issue in international relations theory is which units of analysis and which sources of explanation tell us most about events. While no one level of unit or source of explanation is always dominant in explaining international events, examining interactions among units, particularly durable or recurrent patterns in such interactions, reveals how units are differentiated from one another, how they are arranged in the international system, and how they stand in relation to one another in terms of relative capabilities (Buzan, 1995: 204-5 and 212-3). In thinking about how to define varieties of international regions, it is therefore worth noting the significance of interaction capacity for the different types of interaction (Buzan, Jones, and Little, 1993a: ch. 4 and see Buzan, and Little, 1994: 231-56).

In the first two decades after the Second World War the themes of the regional literature can be classified under four headings. Firstly, the theme of normative writing which debates the problem of world order; second, in response to this mainly speculative material, a number of empirical studies were carried out which attempted to answer the question of whether regional security organizations are, in fact, effective in keeping the peace; third, a constant flow of descriptive materials which either gave accounts of the regional structures in traditional institutional terms, or attempted to deal with particular regions in the multidisciplinary style of the new "area studies" approach which emerged in the late 1950s; fourth, throughout this period there was a separate debate about the respective merits of regional versus universal functional interaction. The latter theme gave rise to the "neofunctionalist" theories which are discussed at length below (Claude, 1959: ch. 6., see also Banks, 1969: 339-40).

Although a review of the extant literature offers a heterogeneous body of knowledge, it is possible to establish a level of coherence when these theories are arranged along a continuum. At the two extremes of the continuum, regionalists focus completely on either the internal or external factors that influence regionalization. These two distinct perspectives are supplemented by a significant body of literature which operates at the centre and at various degrees along the continuum. Theorists in these categories attach different weighting to the significance of internal and external dynamics on region-building (see Figure 2.1).

CULTURAL TRAITS   **INTRUSIVE SECTOR**

Climate   *Language*   *Heritage*   *State System*   *Systemic Factors*   *Geo-political*   *Geography*
                                      Political Structures   *Intl. Org. Membership*

INTERNAL   **EXTERNAL**

Civilizations	Neo-Functionalists	Transactionalists	Double Pincer	Systems Theory	Intrusive Dominant
MacNeill	Deutsch	Wallerstein	Cantori & Spiegel	Kaplan	Vayrynen
Braudel	Russett	Lindberg	Buzan	Brecher	Liska
Toynbee	Etzioni	Scheingold	Rizvi et al	Binder	Hemdel
Huntington	Sundelius	Schmitter	Little	Cobb & Elder	Reinton
Sorokin	Fleming	Nye	Waever	Thompson	Miller
Wright	Claude	Puchala	Archarya	Jervis	Kaiser
Hodgson	Haas	De Vree	Modelski	Krasner	Ayoob
		Mace et al	Wriggins et al	Young	Dominguez
		Keohane	Berton	Zartman	
		Palmer	Banks	MacFarlene	
		Grant et al	Sigler		
		Hettne	Hurrel		
		Brams	Hitti		
		Stubbs & Underhill			
		Wallace			

Figure 2.1 - Spectrum of Literature on Regionalism

One group of scholars who emphasize the importance of internal dynamics when assessing the international politics of regions are those that perceive the world through the lens of civilizations. A recent article by Huntington postulates that nation-states will remain the most powerful actors in global affairs, but the fault lines between civilizations will be the battle lines of the future. He defines a civilization as a cultural entity: villages, regions, ethnic groups, nationalities and religious groups are all characteristics that constitute civilizations. A civilization is thus the highest cultural grouping of people and the broadest level of cultural identity that people have (Huntington, 1993: 22-4).

Huntington is essentially picking up from where other scholars contending with theories of civilization left off. Toynbee, Sorokin, and Wright have all pointed to the authoritarian, egoistic and compulsive nature of civilization as its war-making essence (Toynbee, 1972; Sorokin, 1962; Wright, 1965). Braudel discusses at length the notion of civilizations with particular reference to the Mediterranean in the sixteenth century (Braudel, 1973: 757). He identifies the three great Mediterranean civilizations, Latin, Islamic and Greek, which he describes as groupings of sub-cultures linked by a common destiny.

There are at least three criticisms of this view. First, it seems anachronistic. Before the nationalism of the nineteenth century, peoples felt truly united only by the bonds of religious belief; in other words by civilization. The rise of the nation-state has not replaced, but has certainly diminished this trend. Second, although increased interaction - greater communication and transportation - among peoples has not produced a universal culture, and has on occasion amplified perceived differences among states, it can also be argued that it has assisted in removing some of the misconceptions that previously existed. Third, not enough attention is given to the possibility of restructuring civilization so that human relations are more egalitarian, altruistic and compassionate. Focusing exclusively on the internal features of the international politics of regions leaves little scope for the possibility that modernization and economic development have a homogenizing effect and could produce a near universal civilization structure. This approach suffers from oversimplification, generalization, selective presentation of facts, and historical misinterpretation (Couloumbis and Veremis, 1994: 36-44).

Integration theory, which presupposes the concept of a region, operates completely within the internal dynamic end of the continuum. The first of the two integrationist theories, and the most influential one, is that of the neo-functionalists. Haas, Lindberg, Scheingold, Schmitter, and Nye are among those scholars identified with this viewpoint (Cantori and Spiegal, 1973:

468). One of the weak features of this theory is the apparent insularity neo-functionalists attach to the regional integration process. In other words, their assumptions fail to make theoretical provisions for the influence of extra-regional powers.

A similar theory was presented by Amitai Etzioni in the 1960s (Etzioni, 1965: 220-1, and Etzioni, in James N. Rosenau, ed., 1969: 346-58). He explains the existence of regions and regional institutional co-operation by emphasizing the common cultural 'background variables' as well as data about the internal transactions between the inhabitants, goods, capital and services. Etzioni includes culture, tradition, language, ethnic origin, political structure, and religion among some of the criteria used to identify the 'background variables' and identitive elements.

Even though scholars such as Etzioni, Nye, and Kaiser made note of the role of external actors in later works, they have viewed them as either positive influences upon integration or, at most, as outside threats against which the subordinate system unites (Kaiser, 1968: 84-107; Etzioni, 1965; Etzioni in Rosenau, 1970: 70-71; Nye, 1965: 870-84). Only Nye later makes provisions for external regional powers to actually thwart regional integration or to assume the role of exacerbating conflict (Nye, 1970: 475).

Next along the continuum are the transactionalists who are also concerned with the process of integration. In the 1950s Karl Deutsch argued that cultural interaction can become so intense that a region can become a security community (Deutsch, 1957). According to Deutsch, although the institutional strategies associated with the supranational co-operation do not necessarily exist, common cultural traits can sometimes, in themselves, be strong enough for the region to transcend international anarchy.

Those who focus on the internal dynamics of regional transformation such as Deutsch, concentrate on indicators such as contacts, interchange, and communication between peoples to measure the level of regional integration (Deutsch, 1966). In the case of the Mediterranean, this perspective corresponds with the school of thought that advocates the case for the establishment of a trans-Mediterranean security arrangement. Like their neo-functionalist counterparts, transactionalists are not influenced by the effect external powers can have on the integration process. According to Deutsch and his colleagues, "even where foreign threats were present, their effects were transitory" (Deutsch et al, 1966: 25; see also Puchala, 1968: 38-64).

The literature that emerged shortly after publication of the Deutsch and Haas works is classified as follows by Banks in 1969: i) the macroscopic approach to the identification of regions using aggregate data analysis, led by Bruce Russett, which is examined below; ii) and iii) the "neofunctional" investigations of Ernest Haas and colleagues, and the social communication researchers of Karl Deutsch; iv) the research of Joseph Nye and others which attempt to compare existing integration theories with other theories of international processes; v) the work of Brecher and other scholars which attempts to apply broad categories of systems theory to regional studies (Banks, 1969: 349-50).

Banks also provides the following summary of trends in the literature during the first two decades after the Second World War. "Regionalism and security" was the principal theme during the first post-war decade. This was replaced by "integration" in the second decade and the period of transition between the two, the late 1950s, coincided with the shift to "systems thinking" (ibid.: 337). Bengt Sundelius also stresses that social dynamics are the main determinant in the creation of a region (Sundelius (ed.), 1982: 177-96). He argues that common cultural traits spur regional elites to perceive their periphery in a similar manner. This view is shared by Fleming who stipulates that the stronger the pan-ideology of the regional environment, the more politics will be oriented toward relationships beyond immediate neighbours (Fleming, 1969: 116).

Although the external dimension is brought into the equation, this assessment is still on the internal dynamic wing of the continuum. This perspective may explain why there has been a resurgence of regional rhetoric in the Mediterranean following the collapse of the Cold War. Fearing that the stability offered by the bipolar nature of the international system of states was being replaced by a more anarchic and volatile model, with the European Union gradually increasing its influence in the Mediterranean theatre of operations, Mediterranean countries reorganized their joint activities in the early nineties to at least give the impression that they would not allow outsiders to completely dictate the course of events in the area. In the case of the Mediterranean, one of the features that would figure in the analysis of those focused on the internal factors, would be the similar climate shared throughout the basin that in turn has had an affect on the lifestyles of the inhabitants. Similar agricultural trends would be another feature of investigation. Neo-functionalists also focus on linguistic trends to help them delineate human collectives. In the case of the Mediterranean area, linguistic diversity (Semitic, Greek, Arabic, Latin) only adds to the complexity of attempting to establish regional borders upon these lines.

More recently, a group of scholars who might be labelled neo-transactionalists have responded to the increasing importance of regionalism.

They argue that the sources of regionalism are to be found in the interaction of domestic politics and the changing international economy, and not in structural changes in the distribution of power among states (Busch and Milner, 1994: 273). They base their theory on the fact that current trends in regionalism precede the decline of U.S. hegemony and the collapse of bipolarity. Anderson and Norheim suggest basing the definition of regionalism on continental demarcation lines, and then by subdividing these areas according to the intensity of interaction among the countries on cultural, language, religious and stage-of-development criteria. They emphasise the necessity to include an assessment of historical trade data as this sheds light on the patterns of relations in any particular area (Anderson and Norheim, 1994: 26). Stubbs and Underhill describe the concept of regionalism as having three distinct dimensions. The first concerns the extent to which countries in a definable geographic area share significant historical experiences and contemporary international relations. The second dimension stresses the extent to which these same countries have developed socio-cultural, political, and/or economic linkages that distinguish them from the rest of the global community. The third factor is the extent to which a specific group of geographically proximate countries have developed organizations to regulate crucial aspects of their collective affairs (Stubbs and Underhill, 1994: 331-2). This group of scholars must be credited for identifying the significant role international organizations play in the development of international regions. They have also started to assess the impact of the increasing globalization of the international political economy on regional dynamics (Keohane and Hoffman (eds.), 1991; Palmer, 1991). In his assessment of regionalism in the world economy, Richard Gibb views the resurgence of current regional dynamics as an attempt to promote the theory of free trade on a more restricted geographical basis: "the resurgence of international regionalism is the result rather than the cause of growing tensions within the established economic and social order" (Gibb, 1994: 7-34). Hettne also tends to emphasize the transnational forces of regionalism in his assessment of the regional factor in the post-Cold War international system (Hettne, 1994: 136-7). One criticism against these neo-transactionalists is that they tend to underplay the important influence that intergovernmental international relations still have on the process of regional development. The increasing importance of the political economy of regionalism therefore dominates too many of the observations in this school of thought. Nevertheless, I adopt this transactionalist approach in my framework of analysis in an attempt to assess the influence of transnational forces on regional development in the post-Cold War climate.

At the centre of the continuum is a considerable body of literature which takes into consideration both the internal and external factors that influence regional transformation. In 1961 Modelski had already identified that international regions are the result of changing great power configurations and variable regional integration pressures (Modelski, 1961: 150). Young developed this school of thought when he highlighted the influence that changes in the nature of the international system have upon regional system development. He particularly focused on the demise of colonialism, the general rise in political consciousness and the spread in active nationalism, and the absence of large-scale international war and accompanying polarization (Young, 1969: 341).

The protagonists in this category are Cantori and Spiegal whose study is used as one of the principal yardsticks in this research (Cantori and Spiegal, 1970). Their comparative framework divides all regions, 'subordinate systems', in their terminology, into a core and periphery. They define the core as the grouping which consists of the principal actor/s in a given region. The periphery on the other hand, consists of states that are in some form alienated from the core. They are much more politically, socially and culturally heterogeneous than the latter, and interaction among the periphery is much less frequent than in the core. Cantori and Spiegal also dedicate a significant section of their study to the influence the international system has upon the international politics of regions. They label this component 'the intrusive system' and largely discuss the various ways external actors can influence the course of events within a particular region.

Cantori and Spiegal's theory on the international politics of regions therefore offers a more complex and broader model for the analysis of regional transformation through operationalizing three sets of variables: the core sector, the peripheral sector, and the intrusive system. Although this model may be theoretically attractive, it remains difficult to operationalize its cluster of variables when dealing with multidimensional international regions. The specification of the variables appears confusing, and sometimes overlapping, as it is difficult to determine clearly into which set of variables the data properly fit. In addition, there is little discussion on how the various variables interact with each other - a fundamental weakness given the fluid nature of this unit of analysis.

Following this are what Cantori and Spiegal would call the empirical systems analysts (Cantori and Spiegal, 1973: 480-7). The empirical system approach provides differing perspectives in regional integration from those of the integrationists, mainly due to their concern with finding solutions to the problem of conflict within the existing local balance of power. In such analyses, they consider a subordinate system's political culture, ideology, intensity of involvement with each other, and the role of great powers (Kaplan, 1968).

Empirical systems analysts such as Brecher, Binder, and Zartman, assess both the internal and external dynamics of a region as illustrated in the importance they attach to both the decision making of leaders and political elites as well as the impact extra-regional powers have on regional international relations (Brecher, 1963: 213-35 and Brecher, 1969: 117-39; Binder, 1958: 408-29; Zartman, 1967: 545-64). The empirical systems approach therefore encourages the postulation of particular types of hypotheses that take into consideration both internal and external influences (Cobb and Elder, 1970).

This double pincer analysis, whereby region building and regional transformation is the product of both internal and external dynamics, is further elaborated by a series of authors, namely Ayoob, Buzan and Rizvi et al., Little, Vayrynen, and Waever, throughout the 1980s. To overcome analytical shortcomings when dealing with such a complex concept, this group of scholars concentrate on regions in terms of security relations (Buzan, 1991a: 189; see also MacFarlene, 1985; Jervis, 1982; Krasner, 1983 and 1985).

In his literature on 'security complexes', Buzan argues that any definition of regional security must include both the power relations and the pattern of amity and enmity among states in a given area (Buzan, 1991a: 189). The states that are part of the region and whose interaction fits this pattern, constitute a security complex, defined as "a group of states whose primary security concerns link together sufficiently closely that their national securities cannot realistically be considered apart from one another" (ibid., 190). In his review of regional security, Barry Buzan identifies that some analysts have overcome the difficulty of examining subordinate systems by concentrating on regions in terms of security relations (ibid.). Such an approach is advantageous because investigators do not have to contend with the number of attributes that a total framework approach would entail. It also avoids the obstacles that integrationists confront since this approach is primarily concerned with the security concerns of regional international relations (ibid.).

Regional transformation is therefore the result of shifts in the internal dynamics of the security complex and the external dynamics associated with the global rivalry exercised in the region by external great powers. Acharya provides a vivid example of this model in his assessment of South-East Asian regional dynamics in the post-Cold War era. While Asian regional powers such as China, Japan, or India might attempt to fill the void left by superpower retrenchment, their relative positions will evolve gradually and will be subject to external influences, such as the role of the United States and domestic developments in Russia (Acharya, 1993: 76). Similar analysis can be found in the study of Wriggins et al. on the dynamics of regional politics with a particular focus upon South Asia, the Gulf, the Horn of Africa, and Southeast Asia (Wriggins et al. 1992: 291-5). The main criticism of this approach is that it underplays the domestic context within which security decisions are formulated and implemented. This seems partly due to the fact that the security complex theory has not been discussed in more detail in international relations literature. Moreover, the fact that the initial case studies concerning South Asia, Southeast Asia, Europe and the Middle East, have not been subsequently tested by other specialists in the field has added somewhat to the ambiguity surrounding this paradigm.

At the other end of the continuum, a body of literature gives priority to the external dynamics that influence region building. Cantori and Spiegal refer to this factor as the "intrusive system - the politically significant participation of external actors in the international relations of a subordinate system". Dominguez argues that the hierarchical system of states is recognized by the countries of the peripheries through restrictions on their foreign policy domains by the countries of the centre (Dominguez, 1971: 208). Reinton underlines the dominant role of the intrusive great power in regional international relations by identifying three instances where smaller actors eagerly follow the former's lead: when international conferences take place, when international agreements are negotiated, when decisions are taken to participate in external wars (Reinton, 1967: 354).

Miller follows up on this line of thought by stating that alliance organization is created as the instrumentality for intrusion into subsystems which will certainly influence the behavioural patterns of the region's actors (Miller, 1970: 377). Kaiser even claims that the level of integration which a region is able to achieve is largely determined by the degree of intrusive penetration in the area. The greater the integration, the slighter the chance for participation of the superpowers (Kaiser, 1968: 105-6). The main shortcoming of this group of scholars is not only that they neglect the significance of the internal dynamics of international regions, but they fail to explain why these factors should not be considered. Whereas advocates of the internal dynamic approach regard states, bureaucracies, political parties, trade unions and commercial enterprises as the principal regional actors, the external dynamic school of thought tends to emphasize systemic factors, states, and geography. The former postulates regional divisions based upon cultural factors while the latter are more inclined to stress geopolitical factors. Thus, internal dynamic approaches concentrate focus on the cultural criteria when delineating a region's borders. In contrast, external dynamic approaches discard these factors in favour of natural geopolitical or strategic landmarks such as valleys, mountain ranges and bodies of water.

Military planners often regard regions in these terms. For example, both NATO and the European military alliance, the Western European Union, approach the Mediterranean in separate sectors when compiling their policy planning working documents. The basin is mainly divided into four sectors according to geopolitical criteria: the Southern European periphery (Portugal to Italy), the south-east region (the former Yugoslavia and Turkey-Greece-Cyprus), the Levant (Lebanon to Egypt), and the Maghreb (Libya to Mauritania). Geographical proximity and the possible penetration from external maverick states contribute to this process of delineation.

A review of the literature thus reveals that most of the attempts to explain the existence and nature of regions fall between the internal/external stools of the debate. Although a lack of consensus exists among the literature at the two extremes of the continuum, a certain homogeneous body of literature provides insight into the multidimensional nature of regional transformation. This study attempts to apply this approach to the Mediterranean in an effort to identify what regional processes are currently taking place and whether these developments are significant enough to talk of a resurgence of regionalization in the area.

2.2 Contemporary Theories about Regionalism

2.2.1 The Concept of Regionalism

Most investigations in regional systems analysis have been hindered by the fundamental problem of defining a region. Hence there is a general tendency towards constructing a taxonomy at the expense of explanatory statements about the origins and dynamics of politics in a given area. One direct consequence of this outcome is that many regional studies authors use the "regional system" label to depict the most important international organization in a specific area rather than the overall context of relations among all the members (Banks, 1969: 335-60).

According to *The Dictionary of the Social Sciences*, the term "region" denotes a geographical area which either possesses certain homogeneous characteristics that distinguish it from adjacent areas or other regions, or which serves as a unit of government or administration.

Some of the areas often described as regions exhibit an assortment of homogeneous characteristics. For example, the Middle East combines a common ethnic and cultural background, economic specialization (oil production), and a special position in international politics. The tendency for a specialized culture to co-exist with a specialized economy (particularly in isolated areas - the Maghreb) deserves to be noted. Identifying the limits of such extended regions is one of the purposes of this study (Self, *The Dictionary of Social Sciences*, 1964: 582-5).

This research is concerned with the concept of 'international regions'. It focuses on the various definitions presented in this category. *The Dictionary of the Social Sciences* also emphasizes that the variety and vagueness of the term "region" derives from a tendency to describe as regions any geographical areas of special interest which do not correspond to the areas of states or their main political subdivisions. *The International Encyclopaedia of Social Sciences* defines a region in the international sphere as an homogeneous area with physical and cultural characteristics distinct from those of neighbouring areas. A region also consists of a group of national states possessing a common culture, common political interests, and often a formal organization. It identifies the Benelux nations (Belgium, the Netherlands, and Luxembourg) and the Organisation of American States (OAS which consists of North and South American states, with the exception of Canada) as examples in this category (Vance, 1968: 377-90).

In its historical review of the term regionalism *The Encyclopaedia of Social Sciences* stipulates that the concept is of fairly recent origin and has yet to acquire any accepted precise definition (Hintze, 1953: 208-18). The term was first used by the Provencal poet De Berluc-Perussis but became widely applied in the 1890s. Critics of the concept claim that it promotes separatist tendencies because of its emphasis on particularism. Regionalism is therefore the counter force to any attempts of centralization. Although sometimes linked to nationalism or sectionalism, regionalism is a distinct term in that it recognizes a higher national unity and superior national interests transcending the attachment to the local region.

In terms of a movement, *The Dictionary of Social Sciences* defines regionalism as follows:

a cultural and political entity which seeks to promote and safeguard indigenous culture and foster autonomous political institutions at a regional level; administrative and political units whose primary objective is to create a democraticized and integrated governmental framework at an intermediate level between the state and the local bodies of government; (Self, 1964: 583-4).

The International Encyclopaedia of Social Sciences classifies regional science as a relatively new interdisciplinary branch within the social sciences. It incorporates disciplines as diverse as regional economics, ecology and theoretical geography. "Regional science" implies the intention to apply rigorous techniques of investigation in an attempt to identify regional demarcation lines. It expresses the intent to create theoretical structures and concepts of general applicability (Isard and Reiner, 1968: 382). More specifically, it focuses on the locational dimension to human activities in the context of their institutional structure and purpose and on the significance of this dimension on the understanding of social behaviour.

The regional science analyst's essential task is to identify: i) location, ii) activity magnitude at the location, and iii) the interaction between locations. All of these are regarded as interrelated. Regionalists must therefore distinguish a distinct area and then identify the patterns of interaction within this area. Having conceived a hypothesis, they then have to create a framework of analysis that will enable them to apply their theory. Finally, cut-off points have to be set along a fixed set of criteria. This process will determine the pattern of regionalization that is adopted. A check-list which highlights such questions as: are the criteria used to identify regions exhaustive in nature?; are the locations assigned to regions contiguous?; has any attention been given to the internal dynamics of the subordinate system?, will perhaps facilitate the regionalists' task of finding a coherent and consistent analytical framework for delineating regions. This task can be summarized as making the definition of a "region" operational, i.e. the concept of a region is but an analytical apparatus for separating certain traits perceived as relevant.

A major initial task that the regionalist must contend with is how to prioritize the plethora of criteria linked to the concept of regionalism. For example, if loyalty or patriotism is considered the principal aspect of regionalism, one could perhaps conduct a survey of residents of the presumed border areas asking them which region they regarded themselves as members of or if separateness is to take precedence, physiographic indices could be included. An area's physical relief, ethnic composition, economic development, historic traditions and religious divisions could all be considered for assessment. If political aspects of regionalism are to be stressed, homogeneity of votes in parliaments of an area, or homogeneity in voting patterns at a national level can be taken into account. If interdependence is selected as a main criterion, the nature of communications would be a good indicator. The size criterion is another feature where no consensus exists. If states are taken as the building blocks from which regions are to be constructed then there is clearly a lower limit to the size of a region. Some scholars suggest an alternative approach : the construction of a hierarchy of regions, with several levels of regions and sub-regions. To date, the majority of empirical efforts to delineate regions have focused on those attributes which might be expected to measure homogeneity (see Cantori and Spiegal, 1970: 1-40).

The term "subregion" is often applied to the next component in descending order. It refers to that group of actors who have a less influential role in the international relations of a region. On several occasions analysts substitute the term "subregion" with alternatives that include: subsystem, sub-zone, sub-district and subgrouping (ibid.). In this study the last of the alternatives, "subgrouping", will be used to describe subregional sectors of the international regions being examined, such as the Maghreb and the Levant. The term subset will be used to describe subsections of an international region that do not share enough common characteristics to be classified as an actual subregion, such as Southern Europe.

One approach that can assist in selecting the criteria upon which regional demarcation will take place is to link the criteria to the definition of the region as a functional unit (ibid.). In other words, an area can be delineated upon the basis established in the definition of the term "region". Unless some consensus is achieved on indices for delineation, there will be no agreement on the proper magnitude of a "region".

Unless separateness is used as the principal criterion, identifying the boundaries of various regions will remain a complex task: "It seems to be agreed that regional boundaries are usually indefinite, being zones rather than lines. In the majority of instances, therefore, any boundaries which may be drawn will be necessarily arbitrary" (ibid.). Regional boundaries must hence be regarded as a fluid phenomenon because the world does not in reality break along neatly perforated lines. This appears to be even more the case in the post-Cold War era where the fluidity of international relations makes it easy for a number of countries to be classified in more than one regional grouping. Perhaps regionalists will be consoled by the knowledge that students of general systems theory and scholars seeking to define and examine subsystems of the international system have also had to wrestle with the same boundary delineation problems. No consensus exists on what constitutes the features of an international subordinate system, but geographical contiguity, interaction, and perception of belonging to a distinctive community are often cited as criteria. A boundary criterion of differences in the quality or frequency of cohesion (the degree of homogeneity), communications (the type and level of interaction), level of power (the distribution of power), and the structure of relations (the degree of co-operation and intensity of hostility) are also suggested by regionalist scholars (ibid.).

In *International Regions and the International System: A Study in Political Ecology*, Bruce M. Russett conducts an historical analysis of research into the idea of a region (Russett, 1967). Sociologist Howard Odum was probably the principal researcher in this field of study in the late 1930s. Not content with the traditional and often non-operationally defined geopolitical or geostrategic regions that were used by students of international politics, Odum and his colleague Harry Estil Moore argued that a region had to be more than an area separated from another by barriers, even geographic ones. Although physiographic terms are significant, the importance of a homogeneity of economic and social structure also had to be considered: "this means that it must comprehend both the natural factors and the social factors" (ibid.). Russett identifies a number of the complexities that have vexed both social scientists and students of international relations alike when attempting to conceptualize the notion of a region. First, areas of homogeneity have often been mistakenly regarded as areas of integration. Second, regions are occasionally explicitly defined by interdependence, as areas within which a higher level of mutual dependence exists than in relations with countries outside the area. As a result the definition of a region is then linked to patriotism or nationalism. Third, a region may also be an entity defined by an ad hoc problem. Central and Eastern Europe's principal claim to regional status among Western observers could simply be the threat posed to the whole area by a nationalist Russia. This gives rise to yet another definition, "a device for effecting control" (ibid.: 1-12). The term "Middle East" apparently originated in the late nineteenth century by the British to refer to an area with common implications for its security concerns in the Eastern Mediterranean. Odum and Moore go as far as claiming that a region "provides an economy for the decentralization of political power" (ibid.). An area is thus depicted as a region for administrative convenience.

In an attempt to establish a coherent framework of analysis in this area of study, Russett identifies five types of regional classification groupings, each based on its own set of distinct criterion:

i) Regions of social and cultural homogeneity; that is, regions composed of states which are similar with respect to several kinds of internal attributes;

ii) Regions of states which share similar political attitudes or external behaviour, as identified by the voting positions of national governments in the United Nations;

iii) Regions of political interdependence, where the countries are joined together by a network of supranational or intergovernmental political institutions;

iv) Regions of economic interdependence, as identified by intraregional trade as a proportion of the nation's national income;

v) Regions of geographical proximity (ibid.: 11).

In general terms, there are two quite distinguishable levels at which it is possible to criticize works of this kind. First, Russett devotes such a substantial portion of his research in search of an ontologically correct definition of the term "region" that he does not dedicate his efforts to constructing theoretical statements about regions. As a result, the definition he proposes becomes so inclusive that it is useless for the purpose of analysis. Second, *International Regions and the International System* illustrates the fallacy of puristic induction: the collection of empirical materials as an end in itself and without sufficient theoretical analysis to determine appropriate criteria of selection. Without specifying his analytic purposes in anything but the vaguest terms, Russett fails to carry out the critical task of matching abstract hypotheses with empirical realities. The quantitative data he presents is therefore not presented in the interest of developing theory but appears rather more as a substitute for creative theory-building (Young, 1969: 486-511).

In *The International Politics of Regions: A Comparative Approach*, Cantori and Spiegal provide a more comprehensive interpretation of the term "region" or "subordinate system" in their terminology (Cantori and Spiegal, 1970: 1-7). Admitting to the difficulties and complexities involved in identifying subordinate systems, they formulate their definition after taking the following considerations: every nation-state is a member of only one subordinate system, apart from the most powerful states that are active in a number of regions, such as the United States or France, and a few states which exist on the borderline between two subordinate systems and may be considered to belong to both, for example the so-called buffer states of Turkey and Finland. All regions are geographically delineated - at least in part - by reference to geographical considerations, but social, economic, political, and organizational factors are also relevant. Thus members of a region are proximate, but not necessarily contiguous. Size, however, does not determine the existence of a region. It may consist of one large nation (China), or it may consist of several nations in a relatively compact area (Eastern Europe).

Within the boundaries of a subordinate system, there is a complex interaction between political, social, and geographic factors. The level of this interaction is crucial to defining the limits of a region and will be discussed at length in the analysis of regionalism that follows. For instance, political, social, and geographic boundaries separate the Southern European countries from their Southern Maghreb neighbours. Indigenous political relationships, geographic factors, and social and historical backgrounds help to define a subordinate system. Accordingly, the European continent is more interrelated than an area like the African continent, which is much more politically fragmented. Outside powers play a role in defining a region, in the sense that they can influence the behavioural patterns of certain states within a given area. This was particularly the case in Europe, Central America and South-East Asia during the Cold War when both superpowers sought client states to enhance their geostrategic position. Although geographic boundaries do not change and social factors rarely do, political and ideological factors are fluid. As a result, the identity of a region is both tenuous and dynamic. A comparison of the position in international relations of both Central and Eastern Europe during the Cold War and after illustrates this point (ibid.: 5-6).

After presenting their seven basic generalizations, Cantori and Spiegal propose the following definition for the term "region":

> (It) consists of one state, or of two or more prominent and interacting states which have some common ethnic, linguistic, cultural, social, and historical bonds, and whose sense of identity is sometimes increased by the actions and attitudes of states external to the system (ibid.: 6-7).

The main difficulty with this definition is that it leaves open how strong the bonds between the numerous factors have to be, and whether an area must cover all of these aspects before it can be classified as an international region. As discussed in the literature review, Cantori and Spiegal split a region into three subdivisions in an effort to comprehend more clearly the dynamics that influence such an assemblage. These are the core sector, the peripheral sector, and the intrusive system (ibid.: 20-33).

They define the core sector as a state or a group of states which dominate the international political proceedings within a given region. More often than not it consists of more than one state which possess a shared social, political, and organizational heritage. The identification of a core sector is simplified when a culturally heterogeneous peripheral sector exists. When this is not the case, other factors such as the degree of cohesion, the size of the geographical area and socio-economic factors have to be considered (ibid.: 20). The peripheral sector consists of all those states within a given region which are alienated from the core sector in some degree by social, political, economic, or organizational factors, but which nevertheless play a role in the politics of the subordinate system (ibid.: 22). The peripheral sector is much more culturally, politically, and socially heterogeneous than the core sector, and the level of interaction among the periphery states is much less frequent. An intrusive system consists of the politically significant participation of external powers in the international relations of the subordinate system (ibid.: 25). Although the core and peripheral divisions are an integral part of the regional system, no assessment of the subordinate system will be complete unless it includes all of the actors with an influence in a given area.

Cantori and Spiegal argue that the level of predominance an external power will be able to espouse will vary but can be classified under two main headings: politically significant involvement and politically insignificant involvement (ibid.: 25-6). An external actor becomes a significant player in a region when its actions affect the balance of power of a region. This model of engagement can be expressed by the possession of a colony, economic or military assistance producing an alteration in the balance of power, political or military alliances, or any other actions that would generally be taken by a country indigenous to the region. Great powers and superpowers are most likely to undertake this type of involvement. In contrast, politically insignificant actors are those whose assistance is regularly limited. It often comprises material aid, trade, economic investment and cultural/educational exchange programmes that do not usually result in participation in the balance of power of a region. Secondary powers and middle powers are often the actors which conduct this type of external policy.

In his review of earlier literature on regional subsystems, William R. Thompson attempts to construct a synthesized and standardized definition of the regional subsystem. His review of the regional subsystem literature leads him to distinguishing 21 attributes used to identify the regional subsystem. As there is little consensus in this body of literature, Thompson proposes that the following necessary conditions must be present for a regional subsystem to exist:

i) the actors' pattern of relations or interactions exhibit a particular degree of regularity and intensity to the extent that a change at one point in the subsystem affects the other points;
ii) the actors are generally proximate;
iii) internal and external observers and actors recognize the subsystem as a distinctive area or "theatre of operation";
iv) the subsystem logically consists of at least two and quite probably more actors (Thompson, 1973: 101).

Thompson concludes that one reason why research into this field has so far failed to make significant progress towards development of any coherent theoretical or descriptive framework is because regional subsystem analysts have not made the effort to exchange their findings and conduct joint research projects (ibid.: 115). The main analytical deficiencies that emerge from Thompson's definition is that he pays little attention to the external dimension of regional dynamics, while he remains silent on the type of relations that he would consider when measuring the intensity of interaction among the actors in a given region. His reference to a "theatre of operation" is also open to a variety of interpretations and therefore needs to be further clarified.

Bjorn Hettne refers to the current wave of regionalism as the follow up phase to the neo-functional integration experiment and praxis in the 1970s. The first wave of regionalism collapsed after the slow-down in West European integration and the failure of third world free trade areas. According to Hettne, the current round of regionalism implies a more comprehensive regional co-operation, with stronger emphasis on political dimensions. It is an expression of a post-hegemonic multipolar world where intergovernmental patterns of interaction are increasingly being challenged by transnational forces.

In an attempt to identify the different degrees of "regionness", Hettne distinguishes five levels of regional interaction. The first level is region as a geographical and ecological regional unit, delimited by natural barriers: "Europe from the Atlantic to the Urals". The second level is region as a social system, which implies translocal relations of a social, political, cultural, and economic nature. Relations can be positive or negative, but either way, they constitute some kind of regional complex. Buzan's security complex theory fits into this category. A transformation from a negative to a positive regional complex is seen in the density of international regimes covering the region. The third level is region as organized co-operation in cultural, economic, political, or military fields. In this case a region is defined by the membership of the regional organization in question. The fourth level is region as regional civil society, which emerges when the organizational framework promotes social communications and convergence of values among certain countries. The fifth and final level is region as a historical formation with a distinct identity and actor capability, as well as a certain level of legitimacy. The complex nature of this organizational expression is illustrated by the European integration process. If the European Union becomes a federal organization it would become a region-state. The principal weakness of Hettne's taxonomy of regionalism is that he does not include the influence of external actors in his framework of analysis. There also seems to be a constant degree of overlapping between the levels of regionalism he identifies as is evident if one compares the second and fourth levels or the third and fifth levels. Although he refers to a classification outlining various "degrees of regionness", his analysis sheds little light on the necessary frequency of interaction that is required for a group of countries to qualify as a region.

After sketching the divergent approaches employed to analyse regions it is apparent that no single set of criteria and principles exists among students of regions to define the concept of a region. Nevertheless, it would be fallacious to conclude that the literature's resistance to clear compartmentalization calls the entire exercise of classification into question. The diverse theoretical approaches to regions offer a number of insights into the character of regional transformation that would otherwise be difficult to identify. The various theoretical approaches employed by students of regions in their analytical investigations does not therefore make this school of research inferior or less coherent when compared to other international relations specialisms. On the contrary, the numerous approaches help to comprehend the complex picture of the competing dynamics which are claimed to generate distinct regions. What is essential is that the various perspectives and their concomitant narratives are complementary and not mutually exclusive.

A review of the literature concerning regionalism thus reveals that it is an area modestly explored and therefore should be subject to further investigation. Few works have been devoted to the analysis of this theme as a whole, and no generally accepted theoretical framework for analysis exists. Two factors may account for this poverty in the literature. First, is the underdeveloped state of the discipline of regional analysis among developing countries. Throughout the Cold War, Western scholars were more interested in the role of the superpowers in these areas, and hence the emphasis was placed on the international relations of these countries rather than on their intra-regional relations proper. The second factor is the relative lack of available data in what is a rapidly changing environment where relevant statistics are kept secret as a matter of national security. A number of countries, especially single party states, which can be characterized as closed societies, fall into this category. Statistics are often state-manipulated and the language is permeated by ideological jargon, often incomprehensible except in the context in which it has been used in the past. Moreover, in some instances the data from such countries are not comparable to Western data. Hence, partly because of the relative inaccessibility of non-Western data and partly because theorists are more familiar with the Western world, they have tended to ignore the non-Western world or, even more worrisome, have tried to extrapolate their findings to other areas.

2.3 The Framework of Analysis

Each subgrouping in the Mediterranean, that is, Southern Europe, the Maghreb and the Levant, are subjected to the same structural analysis to ensure coherence and continuity in this thesis.

An important initial task is to define an "international region". In an attempt to include the necessary and sufficient conditions for labelling a collection of states an international region, I adopt and modify a number of the synthesized and standardized definitions offered in the related literature.

One central theme found in the regionalism literature is that which refers to the importance of the "self-awareness" factor. Thompson's contribution is useful but I avoid the attribute that does not appear central to the concept and only adds to the ambiguity surrounding this notion; namely his reference to the necessity for internal and external observers and actors to recognize the international region as a distinctive area or "theatre of operation" (Thompson, 1973: 101). I also refer to Kaiser's definition: "an international region shall be loosely defined as a pattern of relations among basic units in world politics which exhibits a particular degree of regularity and intensity of relations as well as awareness of interdependence among the participating units" (Kaiser, 1968: 86). His suggestion that there must be an awareness of interdependence among the participating units is not adopted on the grounds that a group of states that interact can still be defined as a "region" whether they recognize themselves, or are recognized by other states, as such. I exclude Hellman's criteria of "an awareness of this pattern [relations] among the participating units" for the same reason (Hellman, 1969: 422). However, as a fundamental guide I do take Hellman's general definition of an international region which he outlines as: "the units involved are nations which exist in geographical propinquity and that the subsystem includes forms of interaction such as bilateral trade as well as explicit institutional and legal relations among the regional states" (ibid.).

Further, four of the five types of regional classification groupings that Russett suggests are incorporated: regions of political interdependence and external behaviour, regions of economic interdependence, and regions of geographical proximity. But his reference to regions of social and cultural homogeneity is dismissed as it results in too open-ended a definition and is not regarded as a necessary criteria for the existence of an "international region" (Russett, 1967: 11).

The definition I therefore propose makes provisions for the interaction capacity, defined as the level of transportation, communication, and organization capability among the actors involved (Buzan, Jones and Little, 1993a: ch. 4). It also includes the external dimension more directly. In concluding this analytical exercise, I propose the following definition:

* the states' pattern of co-operative or conflictual relations or interactions exhibit a particular degree of regularity and intensity to the extent that a change in their foreign policy actions have a direct influence on the policy-making of neighbouring actors (this factor will be described as the regional patterns of interaction);
* the states are generally proximate;
* the subsystem consists of two or quite probably more states;
* the influence of external states is taken into consideration.

This attempt to formulate a coherent working definition of an "international region" is also based upon the following considerations: first, while there is no generally accepted academic definition of "region" in the literature in this field, the common sense notion of a geographically continuous or closely related territory is widely though not universally accepted as a starting point (Banks, 1969: 338). Geographic proximity is also the only practical link between the primary emphases of regional studies in each of the two periods of research since World War Two (ibid.: 351; see also Balassa and Bauwens, 1988: 1421-37). Second, any concept of "region" that abandons the notion of physical contiguity as a necessary characteristic opens up the possibility that any entities related to each other with respect to one or more attributes will meet the requirements for consideration as a region (Young, 1969: 487-8). The definition I propose specifically attempts to avoid a situation in which the term "region" is likely to become so inclusive that it is useless. Regions are therefore international subsystems, that is, groups of units within the international system that can be distinguished from the whole by the particular nature or intensity of their interactions with each other (Buzan et al., Working Paper, 1994: 7).

Third, a review of the literature reveals that the search for a satisfactory "essential" definition of the term region is bound to fail. There is no a priori reason to select some criteria rather than others. Inclusion of the infinite criteria that could be incorporated only results in an open-ended definition of this concept (Thompson, 1973: 93). To avoid this predicament, I have opted for the path of nominalism rather than the route of essentialism, for my theoretical analysis (see Popper, 1964: 26-34, for a classic statement of the distinction between essentialism and nominalism). In other words, the definition I propose points out clearly that the concept "international region" is defined in terms of A, B, C, D. The construction of such a concise approach permits me to proceed with the formulation of hypotheses involving the phenomenon of regionalism on a firmer basis than would otherwise be the case.

Fourth, one major methodological problem is the difficulty in drawing boundaries between regions. It is apparent that the decision to assign any one set of boundaries to any one region is an evaluative or political decision. Hence, boundaries are not entirely clear in every case and the cast of significant actors is not beyond debate (Braudel, 1973: 18). In an effort to overcome the grey area surrounding the task of boundary delineation, "the intimacy of interaction among the participating states decreases as, the edge of one region and the start of the next one is approached", is used as a guiding principal.

Fifth, the concept of subregions refers to distinct groups of states within international regions which share a distinct set of common characteristics that separates them from the other members. In the Mediterranean area, the Maghreb and the Levant, both subgroupings of the Middle East international region, best illustrate this concept. In order to conduct a comprehensive review of the states bordering the Mediterranean basin, Southern Europe is also examined as a subregion. Although this group of states lacks commonalties at a bilateral level, it shares an extensive common core of interests at an international level as demonstrated in EU proceedings. (An in-depth analysis of the patterns of interaction within and between the subregions bordering the Mediterranean will take place in chapter four.)

The next fundamental task is the precise description of regional transformation. I propose that the analysis of international regions should focus on the complex interplay and interpenetration of system wide factors and the features that are characteristic of each region. Recognizing the impact of certain global phenomena such as transnationalism, the differences between individual regions and the fluid nature of the international relations of regions, my solution is to construct taxonomical statements about regions that will enable me to comprehend more clearly regional patterns of interaction of both a co-operative and conflictual nature (see Vayrynen for an analysis of various regional types, Vayrynen, 1984: 347-9).

The proposed framework of analysis consists of the following threefold schema: the transnational international region, the intergovernmental international region, and the comprehensive international region. The degree to which a group of states resembles one ideal type of international region more than another is therefore determined by the patterns of interaction taking place between the states in a particular area. For example, if exchanges between states are largely cross-border in nature, such as in the field of off-shore banking, then the transnational international region type classification best outlines the regional contacts taking place. If on the other hand, interaction is government led, as is the case in the Middle East, the intergovernmental regional model epitomises the regional transactions taking place.

Interaction can occur at all levels including political, military, economic and social types and can be either co-operative or conflict defined. The nature of an international region is determined by identifying the dominant pattern of relations within such a subsystem. For example, intergovernmental links across the Western European international region are co-operative dominant, and are complemented by an intricate network of transnational patterns of interaction. Both levels of interaction are represented in one common institution, the European Union (EU), whose role it is to harmonize as much as possible the agenda of both intergovernmental and transnational sectors. On some occasions this has proved no easy case, as during ratification of the Maastricht Treaty, when transnational forces flexed their muscles to block the intergovernmental led campaign.

If regional interaction is both government and transnational dominant, that is, policies are introduced after a consensus has been achieved between the government and the private sector, then the comprehensive international region is the most accurate modality that describes the pattern of regional interaction. The EU Maastricht Treaty ratification process helps illustrate this point. The Treaty was only ratified after certain modifications were introduced in accordance with the desires of the Danish electorate who had rejected the Treaty during their first referendum in June 1992. The interests of governmental officials were thus harmonized with those of non-governmental political pressure groups, a key feature of an embryonic comprehensive international region. A fully functioning comprehensive international region is one where intergovernmental and transnational relations are intense and institutionalized in a single framework. In Western Europe, the EU, together with the Council of Europe and the OSCE, are important vehicles for sustaining co-operation and strengthening the concept of a European international society. If this process of consolidating co-operation continues and results in the creation of a single European currency and common defence policy, Western Europe will strengthen its comprehensive regional characteristics. When transnational and intergovernmental patterns of interaction exist but do not line up, a quasi-comprehensive international region best describes the existing patterns of relations. In such a model, the patterns of interaction required to establish a comprehensive international region are present, but the will to co-ordinate intergovernmental and transnational forces is not.

These three ideal types of international regions are selected to help differentiate among the various kinds of regional phenomena (this approach is submitted by Kaiser, 1968: 84-107, see Figure 2.2). They are ideal types in the Weberian sense - i.e., they do not exist in reality in their pure form. The three models will serve a check-list function in the empirical analysis in an attempt to identify which patterns of relations are active, neutral or non-existent in the Mediterranean area. The principal aim of this exercise will be to detect any increase in the intensity of relations across the basin and whether there are any signs which indicate that the distinct dynamics operating in Southern Europe, the Maghreb and the Levant, are slowly converging. Such a fusion process would substantiate the notion of a Mediterranean international region again re-emerging.

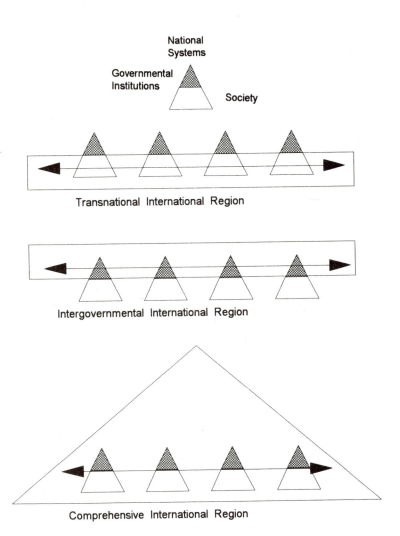

Figure 2.2 - Three Ideal Types of International Regions

These modalities therefore constitute the tools for investigating the criteria chosen in the operational definition of an "international region": they accommodate the interaction between international regions and also help understand the distinct internal dynamics functioning within these regional entities. These patterns of interaction range from the field of "high politics", such as political and military alliances, to economic or cultural relations between the participating societies (Kaiser, 1968: 90).

A systematic analysis of the transnational international region model, the intergovernmental regional model, and the comprehensive international region model now follows. The particular characteristics of each ideal type of international region are outlined and discussed. This critique will assist in the task of identifying the degree to which these ideal types of international regions exist in the Mediterranean area (see chapter four, sections 4.3 and 4.4). The resultant picture will shed light on whether a "Mediterranean region" actually exists, or if references to such a model are a verbal strategy used by politicians to overcome tactfully the boundary zones that divide the area.

2.3.1 The Transnational International Region

The transnational international region model is concerned with cross-border networks of links. States have always been interdependent, but the extension of globalization in the last two decades has led to the situation where about half of the world's business is conducted by transnational corporations (TNCs) (Vernon, 1994: 137-58). These take the form of trade, investment, and capital flows, cross-national corporations, large-sale migration, dense patterns of communication systems, and an instantaneous information structure (Michalak, 1994: 52-7). The phenomenal expansion of transnational interaction, which also includes transnational banking, the emerging process of integrated international production and the increasing importance of off-shore finance, has changed the very nature of the international, and by extension, the regional system (Michalet, 1994: 9). As in any type of international region, this model includes both co-operative and conflictual relations between states.

The decision-makers in this area are non-governmental elites. They pursue their aims by interacting with the social, economic, and political forces which are active in the societies they operate within. These include transnational pressure groups such as Amnesty International and Greenpeace. This category of interaction therefore by-passes the traditional decision-making institutions, such as parliament, but it is nevertheless still influenced by them. In fact, transnational actions often complement governmental policies so as to avoid legalistic clashes with governing officials.

The transnational ideal type of international region assists in comprehending better how a transnational society "connects" states and affects their interaction. An analysis of transnational patterns of interaction will shed light on social movements, economic trends, and even political developments between a particular group of states. This construction also enables us to investigate whether transnational exchanges act as catalysts for a growing interdependence within the international system, including contacts across international regional lines. The question "do transnational international regions stimulate other similar regional configurations?" expands our field of concern from the task of international region formation to that of interaction after such areas are functioning. Unfortunately, little investigation has been conducted in this field of research and it is thus too early to conclude if any specific patterns are observable. Some of the questions which need to be addressed include: how do certain regions react to the elaboration of specific policies of multinational companies vis-à-vis those regions?; does economic interaction between societies in an international region stimulate responses on a regional level elsewhere?

The interaction between transnational international regions at first appears important as this is an area of social action in which one of the constituents of "interdependence" is located. Both economic theory, such as Viner's concepts of "trade creation" and "trade diversion", and communications theory (employed to measure the intensity of relations in trade or in shared membership in international organizations) have registered some of the phenomena in transnational society (Gibb, 1994: 28-32). However, they have not been able to explain their social and political relevance, let alone their causes.

Comprehending a state's domestic features is certainly an asset when examining the transnational regional model. Features such as a state's geographic position, population and social structure, economic capability, military capability and political structure are all relevant. This level of analysis is therefore concerned with the degree of cross-border interaction. Identifying the degree of transnational patterns within an international region and its respective subgroupings of states, facilitates the task of distinguishing any increase in the intensity and/or convergence in regional relations.

Cantori and Spiegal define cohesion as "the degree of similarity or complementarity in the properties of the political entities being considered and the degree of interaction between these units" (Cantori and Spiegal, 1970: 10). A problem with this approach is that Cantori and Spiegal treat the factors of homogeneity and interaction as part and parcel of the same process. In reality, a group of states does not necessarily have to be homogeneous to achieve a high level of interaction. The high level of trade between countries such as the United States and Japan, Europe and the Middle East, and Europe and the Maghreb, demonstrates this fact clearly.

As stated earlier, my primary concern is to identify the degree and pattern of regional interaction within a particular framework. I therefore exclude the homogeneity factor in my definition of an "international region". Any reference to "cohesion" is therefore concerned with the degree of symmetry that exists in state relations. In the context of international regional relations, the symmetry pattern is active when interaction produces similar processes within both regional systems (Kaiser, 1968: 98). In this respect, balance of power theory helps to explain why a certain similarity of behaviour is expected from geographically proximate states. Yet, while expected behaviour is similar, it is not identical (Waltz, 1979: 122).

Parallels can be drawn between the role the concept of transnational interaction plays when considering international regions and the role the concept of integration plays when assessing the nation-state. However, a certain fundamental difference between the two approaches does exist. When applied to the discipline of international relations the concept of integration often implies that the states being assessed will relinquish some of their sovereignty as they become more integrated. No such conjecture is associated with the concept of transnational interaction. The fact that states become more identical and interactive does not mean that they will unite or federate. Different groupings will have arrived at different stages and degrees of cohesion. Examining such a trend over a relatively long period of time will help identify whether solidarity in a particular area is actually increasing, diminishing, or not changing at all. Transnational and intergovernmental connections and disconnections between Southern Europe, the Maghreb and the Levant are discussed in section 4.3 in an attempt to identify any intensity or signs of convergence in their patterns of relations.

Transnational interaction can be assessed from a variety of perspectives. "Social" transnational interaction pertains to the extent of interaction among the inhabitants of an area. The degree of social cohesion varies from one region to another as evident if one compares the higher level of social cohesiveness in Western Europe to the lower level of social cohesion in the Middle East. Transnational interaction at a political, military and economic level can increase or hinder transnational patterns of social interaction. For example, the fact that a government adopts either a restrictive or open immigration policy will certainly affect the pattern and level of migration flows within its jurisdiction. On the other hand, economic indicators also influence demographic trends. Migratory trends in the latter part of the twentieth century have shown that flourishing economies are likely to attract "outsiders" to the area and encourage local citizens to stay. Conversely, a faltering economy will probably have the opposite effect.

Under the rubric of transnational economic interaction, particular attention is paid to the trade relations and the distribution and complementarity of economic resources. In other words, a high degree of economic cohesion is present when a strong network of intra-regional trade is taking place. Western Europe is one example of an international region where horizontal trade is very high. In areas where vertical trade is dominant, that is, the majority of trade is carried out with countries outside the region itself, then economic co-operation and co-ordination is practically non-existent. Trade statistics for the Maghreb reveal such a trend, with approximately only three per cent of the trade conducted being intra-regional (Spencer, 1993: 47).

Transnational political interaction includes all non-governmental political forces that in one way or another attempt to influence political thinking across their region. The nature of the political groups in this sector varies, but some of the more popular organizations are those that contend with environmental or human rights issues. In comparison, cross-border military interaction includes all non-governmental exchanges of a military nature. Purchasing of armaments, the construction of weapon production plants, combat training, and the exchange of strategic planning documents are all types of regional interaction which fall in this category.

Transnational international regions foster various levels of interaction. These include: personal communications (mail, telephone, fax); mass media (newspapers, television, radio); travel (educational exchange programmes, tourism, diplomatic visits); and transportation (rail, road, sea and air traffic) (Miall, 1994: 5). The first three aspects are highly affected by the standard of living and the level of education that have been achieved in an area, while all four are influenced by the physical relief of an area and its level of technological development. Comparing the nature of communications in the United States and Western Europe with those in West Africa for example, reveals the large differences which exist in this field. The promised installation of an information highway by the start of the next century in the United States augurs for a period where certain areas will possess the infrastructure to further promote and sustain transnational flows.

The transnational international region modality is therefore largely concerned with what degree of non-governmental interaction takes place in a given area. Intergovernmental exchanges certainly influence levels of transnational interaction - they can foster, neutralize, or hinder this type of interaction. However, where transnational contacts are dominant among a certain grouping of states, a transnational type of international region has developed. In the Mediterranean area, transnational patterns of interaction are most active in the areas of oil and Islam. Non-state relations in the energy sector are the most advanced north-south co-operative exchanges of this type in the area. In contrast, transnational Islamic forces have largely contributed to conflictual patterns of interaction in south-south relations. In general, non-governmental relations in the Mediterranean remain very limited and there is nothing to indicate that such a network is going to challenge or complement the existing intergovernmental dominant system.

2.3.2 The Intergovernmental International Region

The second ideal type of international region in this taxonomy is the intergovernmental model. In this archetype, decisions are taken and agendas set by elites located in governmental institutions, i.e., the Heads of State and their supportive network, the civil service. This construct can be labelled as relational as it focuses on the orientation of a state's foreign policy behaviour. It covers the complete range of foreign policy options including alliances of a cultural, economic, political and military nature. Like its transnational counterpart, the intergovernmental model reflects the degree and intensity of co-operative and conflictual state relations.

Kal Holsti defines orientation as a state's "general attitudes and commitments toward the external environment, its fundamental strategy for accomplishing its domestic and external objectives, and aspirations and for coping with persistent threats" (Holsti, 1977: 109). This level of analysis therefore deals with both the internal and external dynamics of regionalism. It seeks to address questions such as: what are the country's general objectives and strategy at the global and regional levels?; how do orientations change over time and what are the sources of change? The intergovernmental international regional model helps answer Waltz's question: "what are the relative strengths and foreign policy capabilities of a region's states?" It also assists in understanding Buzan's theory that states' enduring patterns of alignment and conflict are crucial to understanding their relationships (Waltz, 1979: chs. 5 and 6, and Buzan, 1991: ch. 5).

The spectrum of relations among a group of countries in a region can extend from very close co-operation of a bloc to the opposite extreme of direct military confrontation. The degree of amity varies in accordance with the type of interrelationship that exists among a group of states. A bloc is the most intense structure where states act internationally as if they were a single political entity. In an alliance states co-operate with one another in specific ways, while at the tentative co-operation level they co-ordinate policies on a much more limited scale.

At the other extreme of the spectrum, complete enmity results in direct military conflict between regional states. Sustained crises exist when certain states are continuously attempting to shift the balance of power in their favour without reverting to military combat. A stalemate is arrived at when contention continues but the actors involved are aware that they cannot change the situation. They therefore settle for the status quo rather than face the consequences of their actions. At the centre of the spectrum is the equilibrium which signifies a mutually acceptable division of power among the states. The main difference between stalemate and equilibrium is that in the former the states in question would change the situation given the opportunity. In an equilibrium situation neither of the parties wants to alter the balance of power (see diagram below).

bloc>alliance>limited co-operation>equilibrium>stalemate>sustained crisis>direct military conflict

A number of factors influence the course of relations between regional states which makes it difficult to arrive at clear cut conclusions about the nature of any bilateral or multilateral relationships. For example, a state may be in conflict with another state in one sector but have perfectly normal relations in others. United Nation's embargoes help to illustrate this point. While Mediterranean countries are enforcing the trade embargo against Libya in accordance with UN resolutions, they continue to maintain normal diplomatic relations and import Libyan petroleum. It is therefore necessary to distinguish the intensity of conflict and co-operation between particular states in an international region or across international regions to understand better the intergovernmental regional dynamics which are in operation.

The variability of intergovernmental regional relations is complicated by the involvement of external powers. An assessment of international regional interaction concentrates on which states co-ordinate their policies and which adhere to an independent and sometimes aggressive agenda (the spectrum of relations); the basis for their amity and enmity (the cause of relations); and the resources which they employ to effect their relations (the means of relations) (Cantori and Spiegal, 1970: 17).

Further insight into the spectrum of relations within a region is achieved by referring to the resources used in such relations. Is warfare a common means of conducting relations or is the emphasis on implementing co-operative diplomatic relations? A distinction can also be made between those states that are consistent in their foreign policy behaviour and those that favour a maverick external policy. This parameter explores how domestic politics affects the foreign policy behaviour of the major regional states. It also considers how the activities of the major outside powers, operating within the Mediterranean area, affects the patterns of regional interaction. For the sake of clarity, an assessment of the role of external actors is conducted in a separate chapter dedicated to this aspect (see chapter six).

Intergovernmental social interaction includes all those measures adopted by the state that in some manner influence social patterns of contact within a given area. For example, immigration policies have a direct impact on the level of social exchanges that occur. A government's approach to this question can hinder or facilitate migratory trends, depending on how restrictive or open governmental policies are in this field. Intergovernmental economic interaction requires that at the very least states accept and/or develop the basic conditions for international trade: some common units of currency, and legislation that allows goods to cross borders in both directions. At a more co-operative level of economic interaction the parties involved will extend agreed tariff concessions to others participating in the reciprocal system, thus establishing a most-favoured nation network. The spectrum rises through a sequence of stages, each building on the one before it: free trade areas, where quota restrictions and tariff barriers between neighbouring countries are removed, but maintaining their separate restrictive arrangements with the rest of the world; customs unions, where members abolish restrictions on the movement of goods, capital, services and labour; economic union, where participating states harmonize their economic, fiscal, and social policies (Buzan, 1994: 7).

Diplomatic interaction is another type of interaction which falls within this category. The extent of memberships in international organizations assists in identifying the degree to which an international region is institutionalized in a single political framework. For example, a very large set of institutions manage interdependence in Western Europe. This complex web of international institutions serves as a vehicle to strengthen the European international society (Miall, 1994: 5-14). A comparison of international organization memberships across regional divides helps disclose the level of asymmetry in this area of international region interaction. It is thus significant to note to what extent, or not, a regional organization is coterminous with a region's boundaries. For example, compare European Union (EU) or Arab League membership to that of the Organization of American States (OAS). Thus, if all of the members of a region also belong to a regional international organization, this can strengthen the degree of political interaction. This will certainly be the case if the organization's boundaries coincide with that of the specific regional grouping, like that of the Arab Maghreb Union. Membership in a larger organization can therefore serve the dual purpose of harmonizing foreign policy agenda and enhancing the voice of regional actors on the international scene. Although it is by no means a foregone conclusion, overlapping international organization frameworks can help foster regional and international patterns of co-operation. The North Atlantic Co-operation Council (NACC) and the Partnership For Peace (PFP) programme demonstrate such intergovernmental types of interaction. Voting behaviour in the United Nations could also shed some light on the degree of organizational cohesion among certain groupings of states such as the Non-Aligned Movement.

Intergovernmental regional military interaction comprises all types of military exchanges that take place between states in a given area. These include the importing and exporting of armaments, the assistance in training combat units and the construction of military production facilities. Intergovernmental regional relations can be both bilateral or multilateral in nature. For example, Middle Eastern states maintain bilateral military relations with neighbouring Arab countries, like that between Syria and Lebanon, in addition to multilateral contacts through organizations such as the Arab League. In Western Europe, states maintain strong bilateral military contacts, as exemplified in cross-border manoeuvres between countries like France and Belgium. Multilateral ties in this field include memberships in military alliances such as NATO and the WEU.

Assessment of political interaction in international regions is thus concerned with the pattern and degree of complementarity of regimes in the area, and what bearing this will have on the cohesion process. In this context, one can compare the different influences parliamentary-type regimes and autocracies will have on such a process (Cantori and Spiegal, 1970: 12). Across the southern shores of the Mediterranean, the absence or weakness of democracy makes domestic consensus behind governments and regimes remain fragile. In order to obviate this handicap, authorities and regimes adopt pan-Arab or Islamic foreign policies - policies asserting Arab or Islamic interests - in an effort to maintain their positions of authority. In other words, foreign policy is used to alleviate internal tensions and create consensus for the leadership (Dawisha, 1988: 260-75). This form of "regional nationalism" intensifies intra-regional rivalry and is conducive to conflictual international relations. For example, this was the approach adopted by Gamal Abdel Nasser when he assumed power with the support of the Muslim Brotherhood. It was also the Islamic platform of Khomeini when he reigned over Iran, and is nowadays the pan-Arab platform adopted by Saddam Hussein in Iraq.

In the Mediterranean area, the intensity of intergovernmental relations within distinct international regions and across the contiguous international regions, is illustrated by the high degree of cross-national interference and the level of competition between regimes for regional leadership. These two intergovernmental patterns of interaction contribute to the conflictual nature of the relations which characterize the area. The end of the Iran-Iraq war seemed to usher in a period when regional association and collaboration among Arab countries was possible as reflected in the formation of intergovernmental bodies such as the Arab Co-operation Council (ACC) and the Arab Maghreb Union (AMU). However, the 1990-91 Gulf War revealed the fragility of co-operative intergovernmental relations in the Middle East.

As long as the policy of "regional nationalism" is pursued, it will continue to upset efforts to establish an equilibrium based system of relation across the Mediterranean. Regional activism, which aims at undermining the internal political dynamics of the other regimes, and the pursuit of national territorial interests, upsets the balance of power equation. While conducive to the formation of conflict-based regions, it is counter-productive to regional harmonization schemes.

Throughout this taxonomical exercise, it is essential to remember that we are dealing with ideal types of international regions that do not exist in perfect form. Governments can become highly interventionist, decide to form alliances, or organize intergovernmental international conferences, (such as the 1996/97 European Union Intergovernmental Conference), but all of these measures will not always be influenced by intersocietal relations. Alliances such as NATO or the OSCE correspond to this ideal type only if we focus on relations between the participating units along functional lines, that is, in military ventures.

A working definition that comes close to describing this type of international region is that offered by Hurrell:

> ... a set of policies by one or more states designed to promote the emergence of a cohesive regional unit, which dominates the pattern of relations between the states of that region and the rest of the world, and which forms the organising basis for policy within the region across a range of issues (Mace, Belanger and Therien, 1992: 120).

An intergovernmental international region is therefore concerned with what degree of governmental interaction takes place in a given area. The type of interaction can be of a social, economic, political or military nature, but it is always government led. Assessing foreign policy strategies is therefore the clearest way of identifying intergovernmental regional trends. Applying the intergovernmental model will thus assist in deciphering the intensity and patterns of such relations in the Mediterranean area. An attempt is made to see how strong intergovernmental social, economic, political and military ties are and whether any indicators point towards an intensification of such relations across the Mediterranean.

2.3.3 The Comprehensive International Region

The third ideal type of international region is the comprehensive design. This model exists when both transnational and intergovernmental relations are functioning among a certain group of states. Policies and legislation are formulated and executed by both governmental and non-governmental elites with a view to attaining identical, similar, or related objectives (Kaiser, 1968: 93). Decisions in this sector are often taken jointly or in consultation and are therefore of a co-operative nature. The federal measures introduced in the Maastricht Treaty illustrate the type of decision-making process that is characteristic of such an international region. The Western European international region is best described as a quasi comprehensive international region. Transnational and intergovernmental forces associated with a comprehensive entity have continued to evolve across this region in the last few decades.

Political decisions of this kind are those that attempt to reflect the general will of the people, including changes in social trends and attitudes. As such, a comprehensive international region is one where the political elite and their constituents do more than interact regularly. They formulate policy together after discussing the various alternatives available. The Maastricht Treaty ratification process, as discussed earlier, illustrates the type of decision-making process that would become a dominant feature once a comprehensive international region developed.

Social patterns of interaction in this type of international region aim at removing all national immigration barriers and establishing the freedom of movement for all participating states' citizens. The abolition of border controls among European Union member states is an example of such a measure in practice. The degree of interaction at this level therefore increases because it is what both governments and transnational forces advocate.

Comprehensive international regions can be areas where economic interaction on the societal level is stimulated or directed toward specific objectives by joint action of the governments (eg., the European Economic Area), or through co-ordination of national policies with the assistance of supranational institutions (eg., the World Bank). For example, after agreeing to harmonize legislation across state borders and thus open national markets to external competition so they can participate as full members in the development of a comprehensive international region, government representatives will call a referendum or election to confirm that their actions are in line with what transnational elites in the area are seeking.

This international region modality includes any collection of states involved in "integration" schemes. It also includes those groupings aimed at establishing "security communities," in a Deutschian sense, such as NATO or WEU (Deutsch et al., 1957). However, any level of integration must stop short of federation, since attainment of such an objective would no longer qualify the area as an international region. Instead, a new national or quasi-national system would come into existence. Thus the creation of a "United States of Europe" would disqualify the European experiment from this category, as the area in question would become more akin to a national system rather than an international region.

Comprehensive international regions contribute to the formation of similar regional structures if i) such regional models have hurt or are expected to hurt the interests of other regions ("the threat effect"), or if ii) the comprehensive modality is relevant as a model to other regions ("the demonstration effect"). In other words, a comprehensive international region can assist in the creation of other regional types whenever other states feel that their interests are being undermined. One could for example argue that the North American Free Trade Area is largely an initiative aimed at countering the integration efforts being pursued by European Union member states. On the other hand, a comprehensive international region may simply be imitated because it has served as an adequate model for problem-solving. It may have also demonstrated how a group of nations can enhance their international weighting by simply participating in a larger network system. This may help explain the emergence of numerous regional integration schemes over the years that in some manner or other have tried to replicate the European process of integration (Helleiner, 1994: 5).

The comprehensive international region is therefore the most intricate of the three ideal types presented. The harmonization of both public and private objectives is central to the achievement of such an international region. As illustrative examples of this type of international region one can consider some of the goals sought by those who adhere to a federalist vision of Europe. The removal of internal borders and the establishment of freedom of movement demonstrates comprehensive international regional social patterns of interaction. The Maastricht Treaty provisions spelling out economic and fiscal convergence measures and the eventual framing of a single currency reflect the type of economic trends that would exist in such a region. Political interaction would be highly centralized. Yet, each member of the region would retain its right to object to propositions put forward by other members of the same community. Thus political integration, to be achieved through the co-ordination and realization of policies through regular meetings, and not federalism, is the type of political interaction in such a region. Military patterns of interaction resemble those set out in the EU proposal to establish a common foreign and security regional policy. Again, each member state retains the right to act unilaterally to defend their national security, but the overriding goal is for all the members to pool their resources together to safeguard the interests of the comprehensive international region as a whole.

In short, a comprehensive international region is one that aims at harmonizing its transnational and intergovernmental policy-making process. Unlike the previous two international region modalities, interaction within this model is largely co-operative. However, co-operation and co-ordination stops short of establishing a federation, as this would automatically disqualify it from the international region taxonomy. At the very least, a comprehensive international region requires that the states involved mutually recognize their sovereign equality. At the maximal end of this spectrum, the participating states would form a confederation, in which the group of states have become sufficiently similar in structure, ideology, outlook, identity and policy, that they can share some functions of government (Buzan, 1994: 4).

Governmental and non-governmental groups work together to remove as many of the internal barriers within their area as is feasible. Yet, they are simultaneously very careful to keep their agenda in line with public thinking so as not to move further or faster than their citizens are prepared to accept. Comprehensive international regions are therefore unimaginable amongst units that do not share many important social, economic, political and military characteristics.

2.4 Conclusions

Having established an operational definition of "regionalism", and explained three modalities of regionalism, the next task is to apply these ideal type lenses to the Mediterranean area in an attempt to identify the degree to which these ideal types of international regions exist. Immediately a number of key questions arise from this endeavour: What does reality portray - the emergence of an international region? A mosaic of various regional types? A number of distinct small regions? Or a boundary zone between separate regions with politicians employing verbal strategy to overcome this boundary zone?

The coherent and concise definition of an "international region" equips us with the minimum criteria required before the concept of an international region can be defined as a reality. The three ideal types of international regions provide us with the tools we need to understand the patterns of interaction present in such regional configurations (see Figure 2.3). Together, the three ideal types of international regions and the hypotheses associated with the degree of interaction between and within them, help discern more clearly the nature of the area under review. They enable the researcher to comprehend the causation and patterns of regional relations, the level of interaction, and the balance of power dynamics within international regions. (Throughout it must be noted that the "international region" is being regarded as an independent unit of analysis. It possesses its own internal dynamics. In addition, all international regions and their respective subgroupings are influenced by four distinct clusters of actors: the core sector, the semi-peripheral sector, the peripheral sector, and the intrusive sector.)

It is the simultaneity and interaction at various levels that distinguishes regional transformation processes in the contemporary era from those in previous times. For example, decolonization saw the transnational ideal type become much more active, but there was limited development toward progressing to an intergovernmental level of regionalism. The distribution of power remained largely unchanged, with the two superpowers setting the international political agenda. As a result, attempts to establish comprehensive international regions in such places as Africa and Latin America were largely unsuccessful.

Figure 2.3 - Types of International Region Modalities

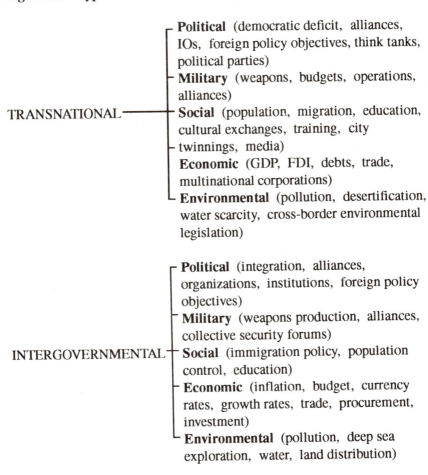

TRANSNATIONAL

Political (democratic deficit, alliances, IOs, foreign policy objectives, think tanks, political parties)

Military (weapons, budgets, operations, alliances)

Social (population, migration, education, cultural exchanges, training, city twinnings, media)

Economic (GDP, FDI, debts, trade, multinational corporations)

Environmental (pollution, desertification, water scarcity, cross-border environmental legislation)

INTERGOVERNMENTAL

Political (integration, alliances, organizations, institutions, foreign policy objectives)

Military (weapons production, alliances, collective security forums)

Social (immigration policy, population control, education)

Economic (inflation, budget, currency rates, growth rates, trade, procurement, investment)

Environmental (pollution, deep sea exploration, water, land distribution)

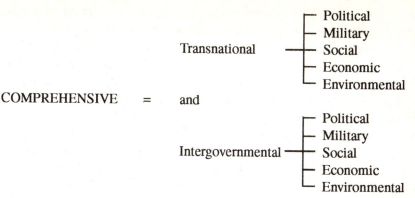

COMPREHENSIVE = and

An assessment of the situation in the post-Cold War period reveals a starkly different picture. Interactions within the three ideal types of regionalism are now characterized by increasing complexity and dynamism. Three distinct but interrelated variables are basic to understanding this development of regional dynamics in the post-Cold War world: i) the change in the structure of the global international order from bipolarity toward a more multipolar structure, ii) new patterns of interaction among the national units, iii) the capabilities and policies of the region's "middle range powers", or so-called "regional power centres" (Vayrynen, 1979: 349-69).

An important advantage in assessing international regions against the broader changes that have occurred in the international system is its comprehensiveness (see Young, 1968). International regions are not only perceived in terms of their relationships vis-à-vis the global system, and their local (bilateral) patterns of international politics, involving conflict as well as co-operation, but in terms of the actions and capabilities of the major regional powers. Only by including an assessment of the major regional actors, can the impact of a change from a bipolar to a more multipolar global international system on Mediterranean regional politics be properly appraised.

At the transnational level, forces are changing the very fabric of the international system. Transnational corporations are acting as agents of that transformation and contributing to the emergence of an intricate network of relations which by-pass the national decision-making framework. Although it may be exaggerating the point to claim that this could lead to the unravelling of the state system as it has existed to date, statistics already indicate that transnational flows are constantly becoming more and more dense and reducing the planet to the equivalent of a "global village". Of central importance in this study is what impact the extension of globalization is having at a regional level? Are developing patterns of transnational links helping or hindering the development of international regions?

At the intergovernmental level, states have been forced to reassess both their global and regional position. The end of East-West rivalry has undermined the status of numerous regional states who could formerly rely on Washington or Moscow for political and financial support (Cuba, Syria, etc.). Even states that decided not to become superpower client states had the alternative of either voicing their concerns as members of the Non-Aligned Movement (India), or else playing the two superpowers off (Vietnam). The post-Cold War era has required regional states to reappraise their foreign policy objectives and strategy to avoid becoming marginalized from the main network of international relations. From a comprehensive regional perspective, the collapse of the Cold War has not only led to a crumbling of the bipolar nature of the international system (a structural change), but has also resulted in a diffusion of power towards the periphery. The implications of this development are far reaching for regional actors. For example, countries such as India, South Africa and Brazil, may regard the implosion of the Soviet Union and the retrenchment of the United States as an opportunity for them to become regional power centres.

Each international region is marked by a uniform, durable set of enmity-amity axes. Often these antagonisms have roots well embedded in bilateral and multilateral relations. Others derive from the manner by which these states were incorporated into the international system during the colonial or Cold War period. One outcome is that the dynamics of domestic politics have become a more prominent feature of regional international politics, often driving foreign policy choices. This development is conducive to an increase in the intensity of interaction in each of the three regional types put forward.

The schema that has been created to examine regional transformation was chosen for a number of reasons. First, it strikes a balance between the two fundamental levels of regional analysis that were described in the literature review: the internal and external dynamic approaches of regional systems. The taxonomy put forward provides an extensive picture of the range of dynamics which can be found operating in an international region. Applying the taxonomy in a check-list fashion to the state relations in any one area will help identify which patterns of interaction are functioning and whether the intensity of relations is strong enough to qualify the area as an international region.

Second, within the confines of a regional system, there is a complex degree of causal relationships and recurrent patterns of interaction. This level of interaction is crucial to defining the limits of a region and to identifying regional transformation. The three ideal types of international regions are an attempt at building theory on the creation and evolution of such regional processes.

Third, between the extremes of parsimony and undisciplined encyclopaedism lies an approach that allows for the construction of viable theory that is neither oblivious to nor overwhelmed by the complexity of regional studies. The suggested framework of analysis is constructed with this basic point in mind. Hence, although the scheme of analysis is not the only one that could be adopted, its principal strengths are twofold: it equips the student of regions with the necessary implements to categorize and illustrate the complex features at work at a regional level, and just as importantly, it is operational.

Fourthly, the proposed framework of analysis was devised with the following check-list in mind:

* are the criteria employed to identify regional interaction exhaustive?
* has sufficient attention been given to the internal, external and structural variants in the contexts to be examined?
* does the approach make provisions for the fluid nature of the regional system?

Focusing the analytical lens on the three international region types that have been outlined assists in the complex process of demarcating and comparing regional entities. The collapse of superpower politics in international affairs and the multiplicity of new states that have emerged in the nineties has again made the 'region' one of the crucial features of international politics. In addition to understanding better the dynamics of international region formation and transformation, this assessment also aims to shed light on how this unit of analysis is being affected by the challenges of the post-Cold War world.

3 An Historical Perspective on Regionalism in the Mediterranean

3.1 The Mediterranean as a Region: An Historical Perspective

As a crossroads of civilizations, the Mediterranean region has witnessed the rise and fall of numerous dominant political forces throughout the ages (Siegfried, 1948: 211-21). The central objective of this chapter is to convey the way in which the Mediterranean basin has evolved as a region. The framework of analysis outlined in the previous section will therefore be applied to the various historical eras, in which leading regional actors continuously challenged one another's authority in the area.

Particular effort is made to examine firstly, the intensity of the interaction within each political-historical era; and further, this aspect is examined in terms of the three ideal-type regional models, outlined in chapter two, within this historical perspective. Hence, the approach adopted in this chapter is one that mixes the outlined analytical scheme with the chronological developments of the area. The purpose of identifying these features is that they can assist in the task of attempting to comprehend better the current phase of regional interaction.

Regional transformations are not evolutionary in that there is no linear process of development. Rather, the process of regional change is largely the result of historical circumstances, that together contribute to the creation of a specific region. A brief look at the evolution of the Pax Romana helps to illustrate this point. One reason the Romans were able to expand their power base at the pace they did, was the lack of external interference in the initial stages of their rise to power. The Romans expelled their more civilized overlords, the Etruscans, in about 510 B.C. and established a Republic. At that time, in the East, the Greeks were more concerned with fending off the early fifth century Persian invasions. In the West, the Carthaginians concentrated on safeguarding their trade monopoly and overseas empire, paying very little attention to central Italy, whose main concern was still agricultural. As a result Rome could extend her power, establishing a single Italian confederacy by 264 B.C. and dominating the whole Mediterranean a little over a century later (Times Atlas World History, 1989: 86).

3.1.1 The Early Mediterranean World: Hellenic Civilization - Roman Empire (3000 B.C. - 565 A.D.): Examples of predominantly comprehensive intergovernmental regional types

The degree of social, economic, and political interaction achieved by the Romans dwarfs that of other states, or groups of states, that subsequently dominated the Mediterranean basin. It can be argued that the Arabs and the Byzantine Greeks each possessed a level of ethnic, linguistic, cultural and historical similarities that together account for their social cohesion. However, neither empire could speak of the same degree of economic or political cohesion that the Romans possessed at the height of their power and neither could co-ordinate the transnational and intergovernmental forces which existed across the Mediterranean.

Italy became politically united when Roman citizenship was extended throughout the peninsula in 90-89 B.C. In 27 B.C. Octavian accepted the title of Augustus and without any serious political rivals as well as the complete allegiance of the armies, he was able to introduce far-reaching reforms throughout the empire, at all levels of society. His reforms included taxation, family and social life, streamlining of domestic and provincial administration to eliminate corruption, and a revival of several old Roman and Italian religious cults. These reforms were supplemented by amendments to the constitution and the setting of external frontiers. Although Augustus kept absolute power in his own hands, he allowed the Senate (the old republican magistrates and business classes) to share with him the task of administering the Empire.

The Roman case is then a classic example of what can be achieved given a range of natural resources and a sophisticated degree of trade within the core and between the core and the peripheries. The pattern of economic interaction on the societal level clearly illustrates a functioning federal system, as all available resources were directed toward the specific objective of safeguarding the Empire.

In the Roman case the conquest of Sicily, which had led to the collapse of Carthage after the Punic Wars, was the making of the Empire. Its large fertile plains and valleys were supplemented by a consistent flow of produce and trade from the North. In fact, Rome was fortunate to possess forests large enough to build her fleets age after age. These resources were augmented by additional sources of iron and timber from places like Corsica and Sardinia after they had been conquered (Rose, 1933: 149).

This example highlights the significance of the land base in creating a strong core in the Mediterranean. History is littered with examples of peoples with a small land base being unable to retain control of their conquests. The Phoenicians, the Greeks and even Carthage, found it difficult to provide a steady supply of manpower and resources to keep their maritime activities going. A look at modern history also reveals that Amalfi, Genoa, Venice, Portugal and the Dutch Netherlands also had a short spell of naval supremacy. Their activities at sea demanded that they throw all of their manpower and economic wealth into naval action. They could only do this so long as the land powers at their rear did not challenge their position. The one great advantage that the Romans had over the modern world was that, in almost every case, they controlled internal free trade. From Gades to Alexandria and the Red Sea there were, in general, none of the customs barriers which have become an integral part of the twentieth century international political economy. Rome encouraged free exchange and, apart from a few short intervals, transnational exchanges flourished. Roman citizens were also free to travel throughout the empire (ibid.: 161).

By contrast, Hellenic civilization lacked the scale of economic and political cohesion that enabled the Romans to take control of the Mediterranean. Even in their victories over Persia, the Greeks found it difficult to maintain a united front. Endless schisms and their rejection of the much required naval reforms led to their eventual defeat, even though their peninsula is better placed to control the Eastern Mediterranean than that of Italy.

Although in theory, the absence of social cohesion can be offset by a degree of economic and organizational cohesion, this is usually not the case. For example, while it can be argued that the Hellenization phenomenon certainly resulted in closer ties and greater unity in the region, it seems that it did not seriously challenge the basically pluralistic character of the Mediterranean. This is illustrated by the fact that for two centuries or more, the Egypt of the Ptolomies and the Seleucid kingdom in Syria, coexisted.

Throughout history different states, or groups of states, formed the central focus of international politics in the Mediterranean. The Minoan and Phoenician seafarers conducted a level of trade that promoted the advance of civilization in the basin. They therefore prepared the route for the Greeks and the Romans, who contributed far more towards promoting a cultural unity throughout the Mediterranean littoral. Indeed it can be stated that between the third century B.C. and the fifth century A.D., the Pax Romana (the longest in duration of its kind), was a federal empire that had a lasting influence in the area. Even when the Roman Empire was challenged and eventually overrun by other empires, the Roman level of transnational and political unity was not to be repeated.

As a strategic centre, Sicily was coveted for its position. It rendered easy the transit between Europe and North Africa, and split the Mediterranean Sea into two roughly equal halves. This too facilitated Rome's objective of co-ordinating cross-border and intergovernmental affairs. Rome skilfully cemented politically active alliances (with Pergamum and Rhodes, for example) which enabled her, though not a maritime state, to defeat all Mediterranean rivals, until the Sea became a *mare nostrum.*

The fluid nature of an international region makes it necessary to continually qualify its delineation. Throughout history, a number of states often linked to a particular international region, by virtue of their activities, move closer to the core or drop out of the semi-peripheral sector altogether. In terms of the three dimensions of regional analysis, this often occurs when a particular state begins to interact more with an extra-regional group of states. As the intensity of interaction increases with this international region, the state can find itself slipping down the hierarchical slope of its own regional grouping and rising up the regional nation-state power hierarchy of its "new" trading partners. In modern history, Turkey's pattern of relations with Western Europe and the Middle East during the late nineteenth century and throughout the twentieth century illustrates the changes that can take place. Although still active in the Middle Eastern international region, Turkey's political, economic, and military ties place it firmly within the semi-periphery of the Western European international region (Anderson and Blackhurst, 1993: 437).

This development is more easily identifiable in the era of navigation, exploration and colonization which began with the arrival of the Phoenicians and Ionians in the seventh century B.C. In the space of two centuries a string of Phoenician colonies were implanted along part of the coast of Sicily, and Sardinia, and along the coast of the Balearic islands and southern Spain, as well as those of Tripolitania and the whole of Numidia in northern Africa. The colony of Carthage, founded in 813 B.C., maintained its allegiance to Tyre until the day when, at the beginning of the seventh century B.C., the metropolis lost its autonomy and the colony had to set itself up as an independent republic. By the early fifth century it was to become the premier maritime power of the western Mediterranean. Throughout Mediterranean history peripheral states played a part in the course of the area's politics, but rarely if ever were they the main protagonists. The colonizing of southern Italy and Sicily (the so-called 'second Greek colonization') which led to the birth of Graecia Magna and, in its wake, rivalry and struggles with Carthage, demonstrates the unpredictable behaviour of these peripheral actors (Ago, 1988: 219).

A review of Mediterranean politics from the sixth century B.C. reflects the dominance of anarchic relations in the area which contrast sharply with the hierarchic nature of relations that were established during the Pax Romana. By the end of sixth century the Greek colonial expansion was complete. The heart of the Mediterranean was occupied by a vast Greek cosmos which, in the west, was confronted by the emerging forces of Carthage and Etruria. In the east it was momentarily confronted by the kingdoms and empires situated in an arc round the eastern Mediterranean, until the immense unitary structure of the Persian empire emerged as a threat. The structure of the Greek cosmos differed greatly from that which prevailed at the time in other parts of the Mediterranean world. It was based on the coexistence of a plurality of poleis, i.e. city-states, each of which constituted a very close knit and jealously independent political formation.

As a result, apart from moments of great peril - such as the Second Medic War in the fifth century B.C., when Persia reached new heights of aggressiveness through an alliance with Carthage, and the Greeks responded by creating a pan-Hellenic league - the history of the internecine struggles between the great Greek city-states of continental Greece, or Graecia Magna, is punctuated with treaties of alliance and military or financial assistance concluded between a city and either Persia or Carthage (ibid.: 223).

Due to the diversity of relations in the Mediterranean through the ages, the task of delineating regional boundaries is an extremely complex task. Several peripheral members are borderline cases, i.e. they only play a role in the politics of a particular region on some occasions. For instance, note the position of Persia which was one of the leading peripheral actors in the Mediterranean during the classical Greek era. Strong historical and cultural elements identify it with the Middle East regional area, whilst in terms of power and the structure of relations, it is certainly appropriate to include it amongst the actors in Mediterranean politics.

A review of this era reveals that all three modalities of regionalism were in a constant state of flux and contributed to the rise of both the Hellenic civilization and the Pax Romana. Analysis of the microparameter demonstrates how the central authorities succeeded in harnessing their human, natural, and economic resources to further their military ambitions. At the cost of some loss of personal liberties, the stable government that the Roman Empire achieved towards the end of the first century B.C., provided two and a half centuries of peace and prosperity. Municipalities throughout the provinces enjoyed a considerable degree of local independence, with the predominantly Latin culture of the west, complementing the Hellenism of the east.

The foreign policy lens of analysis depicts a period of continuous colonial expansion, encouraged by land-hunger, political disaffection, or the desire for adventure or profit. Starting from 334 B.C. Alexander the Great's campaigns east opened up the resources of the Middle East to the Greeks. Even after his mighty empire broke up after his death in 323 B.C., his successor states (Macedon, Ptolemy, Seleucids and Pergamum) continued the diffusion of wealth and a great expansion of trade throughout the eastern and western sectors of the Mediterranean basin. An embryonic network of transnational and intergovernmental relations was therefore established across the Mediterranean (Times Atlas World History, 1989: 76). The Greeks, like the Romans after them, thus executed a foreign policy agenda that saw them not only establish a local community - the polis - but a cosmopolis: a whole civilized and increasingly Hellenized (and later Latinized) world. The Mediterranean area therefore evolved from a comprehensive international region during the Hellenic phase into a fully fledged federal system during Roman times.

3.1.2 The Christian/Muslim World (600 - 1517 A.D.): The intergovernmental modality becomes predominant

When Diocletian came to power in 284 A.D. it was obvious that one ruler could no longer hold the entire Roman Empire together. This is illustrated by his division of power between himself and a joint Augustus with two subordinate Caesars, and his division of the empire into four prefectures and twelve dioceses (Times Atlas World History, 1989; 88). The centre of gravity was gradually shifting eastwards: hence Constantine established a new capital and a Christian city at Byzantium, renamed Constantinople (330 A.D.), while a new taxation system resulted in an economic revival. In any case, further decline followed and although the Empire was theoretically governed by joint rulers, it gradually broke into an eastern and western half, with outlaying provinces falling to barbarian invaders.

Despite this development, the nature of the domestic scene more or less remained the same. For two centuries after the break-up of the Western Empire, the Byzantine monarchy kept Roman institutions and continued to use its Latin courts. Although Greek superseded Latin and the administration was decentralized, it was the Eastern Empire that compiled the two great monuments of Roman law: the codes of Theodosius and of Justinian. The East also preserved and transmitted much of the legacy of the ancient world to the modern one. Even in the West many Roman traditions survived. The Latin tongue, although widely developing into the derivative 'Romance' languages, was still preserved in the Church and as the language of science. Roman law forms the basis of the law of most modern European states and the feats of Roman engineering are still visible in the Mediterranean world.

The network of communications at the time provides further insight into the nature of the regional groupings in terms of personal communications, mass media, interchange among the elite, and transportation facilities. Although the Arabs could never completely replicate the Romans in their external campaigns, they did however succeed in establishing intergovernmental and transnational relations in the Mediterranean area that made them one of the principal actors throughout the eighth century. Further, their naval supremacy was solidified by the conquest of the geostrategic Mediterranean islands which included: Crete, Sicily, Malta, Sardinia, Corsica, Cyprus and the Balearics.

Arab foreign policy strategic planning attempted to somewhat imitate the Roman network of relations. In the period 632-34 they had completed the conquest of southern Palestine and Arabia: by 634 Persia had been overrun (a feat not achieved by Rome) and the next two years saw the Muslims advance to Damascus (Pirenne, 1936: 46). Victory over the Byzantines at the Yarmuk river encouraged them to advance east into Mesopotamia and north-west into Asia Minor. Simultaneously, Arab forces took control of the entire North African shore-line in the seventh century. They then shifted their attention to Iberia, overrunning the Pyrenees in 717 (Koenigsberger, 1987: 114-17). There were further advances into southern France, but the Arab armies were defeated at Pointiers in 732, and in 759 they withdrew south of the Pyrenees. At around the same time they also deployed their forces against the Byzantine empire in the East, besieging Constantinople, and therefore threatening to close the circle of expansion right in the heart of Europe. This expansion owed much to the enthusiasm and religious conviction of the conquerors, but it was also facilitated by the war-weariness of the empires of Persia and Byzantium (Times Atlas World History, 1989: 104).

By the beginning of the ninth century the Byzantium had dramatically shrunk. In the west, the only remaining toeholds were in southern Italy, Sicily and along the Dalmatian coast; Greece was still largely in the hands of barbarian Slavs. Under a fresh and vigorous Macedonian dynasty of emperors, a new era of reconquest and aggressive expansion was embarked upon, and by 1025 the Byzantium frontiers were often close to where they had been in the heyday of Rome.

In 1071 a much weakened Byzantine army was smashed by a force of Seljuk Turks at the battle of Manzikert. In the same year the last Italian possession fell to the Normans; and the period of greatness when Constantinople ruled the wealthiest and best-governed realm in the Christian world, was at an end. The final assault did not come from the traditional enemy, the Muslim infidel, but from the Christian West. It was an already weakened Byzantium which saw the arrival of the first Crusade in 1096, but hopes that Rome and Constantinople could coexist peacefully were soon dashed. Conflictual intergovernmental power politics thus dominated regional relations.

After a series of defeats at the hands of Seljuks and Normans who resumed full-scale frontier aggression in the 1170s, Constantinople itself was seized and ravaged by the swordsmen of the Fourth Crusade in 1204. The main beneficiary was the rising power of Venice, whose fleets had carried the Crusaders. An attempt to set up a Latin Empire of Constantinople proved abortive, as for the first time the Greeks had become a majority within the truncated Empire. In 1261, with the assistance of Genoa, the rival of Venice, they drove out the westerners. The Greek empire was only a shadow of the Byzantium of the past and was no match for the Turks when they advanced into Europe in the fourteenth century.

From the seventh to the eleventh century the Arabs were incontestably the masters of the Mediterranean. Due to their contact with the West and the Far East, the Arabs were valuable commercial intermediaries. The seeds of a transnational type of international region were sown in the Mediterranean at around this time. They imported sugar-cane and cotton into Italy and Africa from India and learned the manufacture of paper from the Chinese without which the invention of printing would have been valueless (Pirenne, 1936: 49). Trans-Mediterranean maritime traffic between the Christian West and the Byzantine and Muslim worlds was at a high in the tenth and eleventh centuries, although the total size of the traffic was certainly minuscule compared to later figures of trade. The fact that the Arabs were both richer and better equipped than their European neighbours added to the conflictual nature of relations in the area. Arab transnational exchanges extended from the Straits of Gibraltar to the Indian Ocean, through the Egyptian ports, which were in communication with the Red Sea, and Syrian ports, which gave access to the caravan route to Baghdad and the Persian Gulf. In contrast, navigation of the Christian peoples was restricted to the coastline of the Adriatic and southern Italy (ibid.: 50).

Commercial interaction intensified from the twelfth century onwards and led to co-operation occurring in the cultural, scientific, and artistic fields. The commercial fleets and navies of the Italians allowed and facilitated this cross-cultural process. Their defensive assistance was sought by all coastline states in the vicinity. In return, the Italian city-states were rewarded by means of fiscal and mercantile concessions.

The decline of the Italian city-states' commercial hegemony in the basin resulted from a series of developments that slowly saw European nation-states gain mercantile supremacy. The discovery of America (1492) and the Cape Route to India, the West Indies and China (1497-98), convinced Portugal, followed by Britain and the Netherlands, to increase their emphasis on maritime undertakings. The opening of this new trade route also meant that an alternative course to conduct trade and commerce between Europe and Asia now existed which was to deal a severe blow to the Mediterranean economy in general. The large territorial base and considerable pool of resources which these European nation-states could tap into soon came to play a crucial role in the evolution of Mediterranean affairs.

The simultaneous successful military campaigns of the Ottoman Turks undermined the influence of Venice and the other Italian cities, as routes to the Asian markets fell into the hands of the enemy. It was only a matter of time before the Italian states system was overrun by more powerful and resourceful actors in the Mediterranean.

Structural fluctuations in the Christian/Muslim world

By the end of the eighth century, the Mediterranean came to serve as a boundary zone between the Christian North and the Muslim South. This balance of power, along religious lines, was predominant in the region throughout the following two centuries. For centuries Europe had gravitated about the Mediterranean. It was by means of the Mediterranean that the various parts of the civilized world had communicated with one another. Under the impact of Islam this unity was abruptly shattered. The intercourse between the East and the West, which had hitherto been carried on by means of this sea, was interrupted. For the first time since the formation of the Roman Empire Western Europe was isolated from the rest of the world (ibid.: 50-52).

Yet, not all states fitted neatly into this arrangement. Although the majority of states on the Northern shore, particularly the Byzantines and the Franks, regarded their survival as tantamount to preventing the progression of the Arab hegemonic campaign, a number of states adhered to an alternative policy. Fearing complete control by the Papacy, and recognizing the potential economic benefits that were available, states such as Naples, Amalfi, Gaeta, and Salerno played down the importance of religious differences, often aligning themselves with the Arabs. Hence, despite the lack of common characteristics, this group of Italian city-states formed an informal alliance by serving as an economic and political buffer between the two primary and contentious forces.

Transnational relations evolved further during this period as a result of the Christian West's succession of crusades to the Holy Land. Christian soldiers were accompanied by merchants and artisans who moved to the Levant, settled down and established prosperous ventures stretching from the Black Sea to the Straits of Gibraltar. The Italian colonies which cropped up throughout the Middle East - Aleppo, Beirut, Damascus, Cyprus, Jaffa, Jerusalem and Antioch - demonstrate a return of transnational forces across the eastern sector of the Mediterranean.

The first part of the Christian-Muslim phase of Mediterranean history therefore demonstrates a shift from a boundary zone management situation to an increasingly transnational Mediterranean area. Conflictual intergovernmental dynamics gained in prominence as the Christians and Muslims sought to maintain and increase their power bases in the Mediterranean basin. The spectrum of relations during the later stages of this period is one of practically complete enmity between the two main poles of power in the vicinity and explains the high rate of direct military confrontation throughout the Mediterranean. The Mediterranean had therefore developed from an outer zone of suzerainty of the Roman empire to a frontier area between the Christians and Muslims, and, as patterns of interaction intensified across the area, into a conflict defined region in which co-operative transnational links were present.

3.1.3 The Rise of the European Nation-States (1500 - 1900): Intrusive intergovernmental regional politics

By the start of the sixteenth century, the Mediterranean basin was already an area controlled by outside powers. Indigenous actors gradually found themselves superseded by the rise of the European nation-states. The significance that the nature of communications can have upon the effectiveness of an external power's desire to influence political relations in a region becomes clear when examining Britain's role in the Mediterranean.

Britain's ability to form a central focus of international politics in the Mediterranean in the seventeenth century is partly due to the impressive chain of command that it established. The British set up their first permanent base in the Mediterranean in 1704 when, during the War of the Spanish Succession, they succeeded in capturing Gibraltar.

Britain's successful campaigns against the French, most notably that at Akoubir (1798), and at Malta (1800), increased Britain's manoeuvrability in both the Central and Eastern zones of the basin (Times Atlas World History, 1989: 187). Its subsequent clashes with the Bourbons in Sicily and the Savoys in Sardinia helped halt a French initiative to create a Napoleonic system of client states. Britain leased Cyprus from Turkey in 1878 and took over Alexandria in 1882. This control of a series of stepping-stones enhanced Britain's intergovernmental relations throughout the basin and enabled it to monitor closely potential rivals' activities. It also explains how the British island state could maintain its dominant position so far away from home. Britain's comparative advantage throughout was its supremacy at sea and it sought to safeguard this at all costs (Kennedy, 1988: 193-203).

A high level of intergovernmental economic and political interaction was achieved when the British, with a security interest in preserving equilibrium on the European continent, perfected balance of power diplomacy. Continuous supplies of produce and war-time materials were obtained from the vast network of colonies that stretched from within the Mediterranean to the key outpost of India. The fact that the British employed these resources effectively, and were successful in replenishing them, largely accounts for their ability to establish a vast empire. Moreover, Britain succeeded in commanding and controlling the area by dividing the basin into two sectors, the East and the West, just as the Romans had done.

Responsibility for these sectors was delegated to allies within geographical proximity. In the West, Britain thus relied on Austria and Savoy to help monitor the naval movements of both France and Spain. In the East,

Britain's main priority was to protect trade routes to India and trusted the Ottoman Empire to keep the Russians in check (Times Atlas World History, 1989: 250-51). However it cannot be argued that Britain's high degree of organizational and economic cohesion compensated for the lack of social cohesion in the area. Even in those colonies where it took complete command of political relations, the lack of social cohesion was never completely overcome - something which was revealed in the subsequent decolonization process in the twentieth century.

Even at the height of British colonial power in the Mediterranean (1815), the cornerstone of London's defence policy remained one of preventing any of the core or peripheral actors from translating their potential power into a rival maritime hegemon. France's continental expansion was already severely limited and British superiority at sea meant that they would have to allocate a large proportion of their resources to maritime programmes if they seriously wanted to challenge the British. Spain also had to accept the abandonment of any Mediterranean endeavours after a succession of decisive defeats at the hands of both the British (Capo Passaro in 1708) and the Austrians (Milan, Naples, and Sardinia at the turn of the eighteenth century).

Russia and Austria were another two external powers whose interests and potential in the Mediterranean gradually increased at the end of the eighteenth century. Russia's primary concern was to gain access and control to a warm water port, which, unlike Russia's outlets on the Baltic Sea, was free from hostile powers. Influence in the Mediterranean was also seen as conducive to Peter the Great's desire to Europeanize his country. Russia's policy towards the Turks was thus regarded as justifiable for ethnic and religious reasons: it saw itself as the rightful protector of the Slavic peoples and as the self-professed heir to the Byzantine empire. As the guardian of Christianity, Russia advanced southward in 1774, at first establishing footholds in the sea of Azov and the Black sea. In 1774 the Czar took command of the Crimean Turkish outpost, with the Bosphorous Straits forced open to its commercial navigation.

Austria's interest in the region increased once it became apparent that the provisions provided for in the Treaty of Westphalia in 1648 would prohibit it from attempting to unify the Habsburg empire under its authority. Realizing that its continental ambitions had received a serious set-back, Vienna concentrated its resources on the Ottoman Turks in the South. Austrian campaigns proved successful in the long-term as the Ottoman empire was eroded.

International relations in the Mediterranean in the second half of the eighteenth and first half of the nineteenth centuries are particularly indicative of the effects the intrusive system can have on the level of power and structure of relations of an international region. Constraints in continental expansion shifted European great power attention to the Mediterranean. As European manoeuvring in this "sphere of influence" increased, the Mediterranean became their geopolitical playing field. When viewed from the capitals of Europe, the Mediterranean was thus assessed in regional terms.

Britain's central concern in the case of Russia was that the Czar might attempt a ransacking of Constantinople in an effort to establish an outlet into the Mediterranean. Such a development would serve a double blow to British interests in the area. First, it could potentially interrupt British commercial ventures in the Eastern Mediterranean. Second, it could also tilt the balance of power in the region away from the British. In other words, if the Russians proved successful in their Mediterranean campaign they would probably join forces with the French and Spanish to overcome British supremacy.

In any case, London decided to dedicate the majority of its supplies and reserves in the Mediterranean to monitoring the French. Revolutionary France restored the traditional expansionist foreign policy agenda of the Bourbons. Its objective was to carry out a process of boundary delineation by verifying its natural boundaries on the Rhine and the Alps. France also began to exert pressure on Italy as a first step towards enhancing its position in the Mediterranean and the Levant.

French ambitions to disrupt British interests throughout the Mediterranean basin failed to get off the ground. They suffered a series of defeats at the hands of the British, most notably that at Aboukir (1798), and at Malta (1800), which restricted France's manoeuvrability in both the Central and Eastern zones of the isthmus. Once again British superiority at sea was the key factor which enabled Britain to resist the French hegemonic undertaking.

Britain maintained its power monopoly after further successful campaigns against the Bourbons in Sicily and the Savoys in Sardinia. Its influence on the structure of relations in the region is well illustrated by Britain's blockade of a French attempt to create a Napoleonic system of client states with Spain and Turkey. Britain achieved this by forcing open the Bosphorous Straits, conquering the Ionian islands and launching a succession of maritime expeditions against Spain.

Over all, change in the Mediterranean's balance of power was minimal between the Congress of Vienna and the latter part of the nineteenth century. Britain retained its dominant position in the littoral, concentrating more and more of its resources to the Eastern sector of the basin.

There are two reasons for this shift in policy. First, Britain became more preoccupied with the Eastern zone because of its ever increasing commercial activities in the Middle and Far East: suffice to say that sales within the Mediterranean were less important than sales beyond it, for all the countries of the littoral combined did not purchase as many British goods as did India on its own (Monroe, 1938: 17). Second, the British wanted to make sure that the vacuum being created by the declining Ottoman empire was not completely filled by the Russians who were relentlessly waiting in the wings.

In the Western sector, Britain's attention was focused upon the containment of France. After its losses in the Americas, the French hoped that they could compensate by territorial gains in the Mediterranean. But Britain's system of communications, level of power and structure of relations were strong enough to prevent this from occurring.

European power politics in the Mediterranean

Cantori and Spiegal describe two types of external regional participation: politically insignificant involvement and politically significant involvement (Cantori and Spiegal, 1970: 25). The former classification refers to those types of measures that are unlikely to influence the balance of power within an international region. Cultural and educational exchange programmes, material aid and trade are some of the actions that fall under this umbrella.

The latter classification refers to all those circumstances where the influence of an outside actor will alter the balance of power within a particular region. There are a number of ways that an external actor can have a direct impact on the evolution of an international region. The possession of a colony, direct military or financial investment, formal alliances, or direct military intervention are just some of the options open to an external actor when attempting to gain a foothold in a specific area. It is this type of involvement that can have a direct impact on the dynamics of region building and thus the type of involvement that this research is primarily concerned with.

The British experience in the Mediterranean is certainly a clear example of politically significant intrusive action because, unlike its predecessors who controlled the basin, Britain was not a Mediterranean nation. As a non-Mediterranean country, Britain's interests in the region were strictly maritime. Like the Roman episode, their dominance is also an example of overlay being realized in the Mediterranean.

The political potential of the external powers in the international relations of a particular region are in a constant state of flux. For example, Britain's position in the seventeenth and eighteenth centuries may be contrasted with the more evenly balanced nature of the intrusive system during the first quarter of the twentieth century. The degree of supremacy which the British had previously held in the Mediterranean basin gradually diminished as other European states, in particular Germany, France, Russia and Italy, increased their influence especially with the demise of the Ottoman empire (Time Atlas World History, 1989: 251).

Realizing the intrinsic decay of Ottoman power, the Egyptian leader Muhammad Ali embarked upon a delicate diplomatic balance between England and France, by playing one side against the other (Vatikiotis, 1969: 68-73). By 1839 it was apparent that the Pasha's military forces were far superior than Turkish military power. King Louis Phillipe withdrew his support for Muhammad Ali and formed a coalition with England to try and save the independence and integrity of the Ottoman State. The British bombardment of the Pasha's positions in Beirut and Acre was a prelude to the London Treaty of 1840 (Brown, 1984: 38-57). Until this point France had been the prime protector of the Muhammad Ali dynasty, with Britain the champion of Turkish suzerainty over Egypt. Anglo-French action against the Viceroy of Egypt led to closer relations between England and Egypt, which in turn led to British domination by the end of the century (ibid.).

The scramble for colonies by the European nation-states and the rise of nationalism are the main characteristics of Mediterranean politics in the second half of the nineteenth century. Both had a significant impact on the nature of the intrusive system in the Mediterranean, which saw more of an equilibrium being achieved among the European great powers' interests. The French set up stepping-stones on the southern shores of the basin by seizing Algeria in 1830 and Tunisia in 1882. The British balanced French progression by strengthening their position of authority in Egypt in the same year. This permitted the British to control the recently completed Suez Canal (1869) which was of very important strategic and economic value (Monroe, 1938: 71-137).

The first traces of nationalism emerged in the Mediterranean in the Balkans, when Serbia and Greece won independence in the 1820s. Italy followed suit after its process of unification in 1869-70. The intrusive system reacted to this development by condoning the region's indigenous course of events, demonstrated by Britain's wholehearted support of Italian statehood. These links were further improved when the new Italian state sought to imitate the days of the Roman empire by acquiring colonies for itself at the expense of the French. Apart from disputing Tunisia with the French, Italy took control of Eritrea in 1885 and captured both Libya and the Dodecanese in 1911-12. Italy's Mediterranean ventures were balanced by the necessity to nurture continental ties, especially in the Balkans, in an effort to bolster its national security (ibid.: 139-203).

These internal shifts within the Mediterranean were complemented by a series of changes that took place in the region's intrusive system. British and other European nations' efforts to prop up the Turkish empire were unsuccessful, and the "Sick Man of Europe" could not regain any of its previous regional power. Russian ambitions in the Balkans were consistently held back by diplomatic and military operations orchestrated by Britain, who wanted to obstruct Russian expansionist plans in the Mediterranean at all costs. The Balkan Wars of 1912-13, however, saw the decaying Ottoman empire regress to Adrianople, leaving the Balkan states formally independent, but severely vulnerable to Russian and Austrian aspirations (ibid.: 207-21).

3.1.4 The Twentieth Century

The emergence of Germany as a leading actor

The late nineteenth and early twentieth centuries witnessed the arrival of another prominent non-Mediterranean actor in the area: Germany. Berlin took advantage of the demise of the Ottoman empire by increasing its influence via the commercial and diplomatic ties with the recently established independent Balkan states. One major project conducted by the Germans was the construction of the Baghdad railway which strengthened their interests in the Levant. Germany also entered the Western sector of the sea, by becoming involved in the Moroccan issue against the French (Monroe, 1938: 231-46).

The emergence of Germany in the Mediterranean displeased all the other actors who already had influence in the region. Britain was already familiar with the Germans' ability to build-up naval forces fairly rapidly, as it had done in the North. In fact, the first four decades after its unification saw Germany amass large fleets and arsenals of weaponry that escalated the level of tension among its regional counterparts. Realizing that it had to counter German projections in the South early on, if its position of supremacy was not to be undermined, Britain exploited France's similar concerns regarding Germany and signed a Franco-British solidarity pact against the Kaiser.

In 1879, Bismarck sympathized with Russia's efforts to consolidate its position in Bulgaria, and even exploited Germany's new role as an ally to force Austria-Hungary into line. He poured scorn on Austro-Hungarian attempts to enlist the support of Britain, Italy and Germany against Russia. Rebuffed by Gladstone, Austria-Hungary fell back on co-operation with Germany and Russia in the Three Emperors' Alliance in 1881. This Alliance survived the crisis over the union of Bulgaria and East Romania in 1885, but the Austro-Russian contest for control of Bulgaria (1886-7) destroyed it. Bismarck had promised the Russians his continued support in the Reinsurance Treaty, but the 'Mediterranean agreements' of February, March, and December 1887 between Britain, Italy and Austria-Hungary (Spain acceding in May) to resist supposed French and Russian designs in the Mediterranean and at the Straits, annihilated Russian influence in Bulgaria (Times Atlas World History, 1989: 250). Italy acceded to the Austro-German-Romanian alliance in 1888, and between 1889 and 1894 Germany, with a new emperor and chancellor, swung into line behind the Mediterranean Entente. In the meantime, France drew steadily closer to Russia: the first of a series of loans was concluded in 1888 and a military convention was signed in 1894. In 1897 the Germans had abandoned the 'New Course' ceasing to underwrite Austria-Hungary in the Balkans and co-operating in the Far East with Russia and France. Fear of German hegemony in continental Europe and across the Mediterranean persuaded Britain to settle its long-standing colonial differences with Russia in 1907, and together with the French, they later formed the Triple Entente. Italy also relaxed its ties with both Germany and Austria after realizing that its interests in the Mediterranean were not being safeguarded through such alliances. Sharing similar ambitions with Britain and France over the future of Libya, Egypt and Morocco, Italy sought a policy of rapprochement towards London and Paris to make up for the loss of German and Austrian support.

At the end of the nineteenth and start of the twentieth century, intrusive system relations in the Mediterranean shifted significantly. The Mediterranean became even more of a conflict based zone as intrusive actors aligned themselves according to the already divisive patterns of relations among the indigenous states. This continued in the run up to and during the course of the First World War. As the number of external actors with an interest in the Mediterranean increased, the degree of competition intensified. The division of Europe into two rival alliances (France, Britain and Russia against Germany, Austria-Hungary and Italy) resulted in a five year conflict between 1914-1918 which weakened further the local powers of the region. The external powers were still able to exercise a large measure of control over the area after the War, mainly due to the large number of colonies and micro-states which they possessed.

The close relationship between transnational and intergovernmental forces discussed earlier in the Roman case-study becomes even more apparent when compared to the much less co-ordinated system that existed at the end of the First World War. The Treaty of Versailles in 1919 did very little to quell disputes and tensions rampant in the Mediterranean area. Diplomatic channels of communication were not powerful enough to cope with the ramifications of the Ottoman Empire's collapse. As a result, great powers found themselves at loggerheads over territorial and political control of the Mediterranean basin. One consequence of this was that the European powers were somewhat taken by surprise by an incipient, but significant, Arab nationalism from Asia Minor to Algeria. Ideas of autonomy, encouraged by the revival of Arabic literature and campaigns to reform the language, gradually gained currency, particularly during the reign of Abdul-Hamid II (1876-1909) (Times Atlas World History, 1989: 229).

In addition, the flexibility of the indigenous countries was also restricted because the victorious external powers in the basin sought to co-operate with each other. While revolutions swept through Russia, Germany and Italy, the centuries-old Austro-Hungarian and Ottoman empires crumbled into collections of smaller new states. By 1925, Britain felt reassured that its position in the Eastern sector of the Mediterranean was as secure as ever. The signing of a treaty with Egypt guaranteeing access to the Suez Canal and the termination of hostilities between Greece and Turkey were two factors that promoted this positive outlook. The fact that Italy controlled Libya, France managed Syria, and the British were granted a mandate in Palestine, convinced London that its access to oil and rubber in the area, and its commercial links with India and Australasia were stable for the foreseeable future.

The maverick behaviour of Italy soon proved that Britain's optimism was premature. Disputes and tensions among the intrusive system's participants again increased and was reflected in the tense conditions within the Mediterranean area. Regional strife soon took over. Uncomfortable with the status quo, fascist Italy launched an ambitious programme of Mediterranean expansion that constituted a threat to the security of many of the regional powers. Italy attempted to pressurize Greece into invading Yugoslavia and Albania, in the hope that it could then use the area as a base to launch further military expeditions into the Levant. In 1935 Italy invaded Abyssinia and then went to the rescue of Franco in Spain, thus gaining footholds in two strategically important locations. Mussolini's next major conquest was his seizure of Albania in 1939.

The six year Second World War again plunged the Mediterranean into a period of fierce rivalry and destruction, with Britain, and later the United States, employing all of their available resources to prevent the Mediterranean from falling to the axis powers. Not only did Europe experience its last hegemonic war, but it also endured an episode that in many ways resembled previous hegemonic struggles. Parallels can for instance be drawn with features of the Napoleonic campaigns: rapid continental victories, efforts to invade Britain, and, once this failed, an attempt to push eastward, against Russia. In both cases, mastery of the Mediterranean emerged as one of the decisive factors in thwarting imperial design. The one significant difference between the Second World War and previous hegemonic wars in the region, was that in 1945 both Europe and the Mediterranean were so exhausted from the campaign that non-Mediterranean powers found themselves in command and control of the security of both continental Europe and the Mediterranean.

The Cold War

The United States and Soviet Union were not the only external powers which influenced the intrusive system. Secondary and middle powers also pursued their interests and therefore also affected the balances in both the intrusive core and Mediterranean as a whole. Britain, for example, still found itself able to play a role in the security of the region, even if it was a rapidly diminishing one. In spite of its declaration in 1947 of no longer being in a position to guarantee the security of the Eastern Mediterranean, Britain retained control of a number of strategically significant bases which included Cyprus, Malta, and Gibraltar.

France also retained an interest with its governing of Algeria, Tunisia and Morocco. In contrast, Italy's military defeat in the Second World War led the Italians to focus on putting their own house in order and thus resulted in Rome's withdrawal from Mediterranean affairs. The involvement of these and other secondary and middle powers in the region initially weakened co-operation among the main actors in the intrusive system. But this predicament was soon to change.

Throughout the forty-five years of superpower rivalry, the Mediterranean sustained a series of developments that affected the internal dynamics of the region. In the first decade of the bipolar international system, the Mediterranean testified to the commencement and bolstering of the United

States' commitment to contain Soviet expansion in the littoral. It sought to realise this policy both directly, by establishing bases in the basin, and indirectly, through political parties which challenged the Communists whose support increased dramatically in countries such as Italy and Greece.

No assessment of foreign policy orientation during the Cold War would be complete unless it included an analysis of the alliance network that developed. After the Suez fiasco in 1956, which saw the diminution of both French and British interests in the area, the two superpowers had a monopoly over the intrusive system. They dominated international regional politics in all global relations, including the Mediterranean. After the breakdown of communications in the Cuban missile crisis in 1962, the United States and Soviet Union gradually built an elaborate global communications system that enabled them to penetrate the peripheral sector further. Superpower politics thus gave rise to an intricate network of patron-client relationships in the Mediterranean. Bilateral contacts were nurtured through the string of client states that both Moscow and Washington fostered in the basin, and encouraged interchange among the region's elite.

Advances in land, sea and air transportation facilities, and the progress registered by the mass media, also influenced the nature of communications around and across the Mediterranean area by helping the intrusive sector maintain its grip on regional affairs. (The implications of the subsequent collapse of the Cold War on this alliance network is one of the topics discussed in chapter four.)

During the Cold War the two superpowers were unwilling to establish a consortium in the Mediterranean and unable to resort to direct confrontation because of the danger of a nuclear holocaust. As a result, three trends of Soviet-American interaction are discernible: mutual non-involvement, unilateral abstention, and restrained mutual involvement (Cantori and Spiegal, 1973: 34). Mutual non-involvement means that neither superpower participates in a politically direct manner in the area or state under assessment. Although mutual non-involvement was not adopted in the basin as a whole, the superpowers did follow this approach within peripheral sectors of the region. This was for example the case in relations with countries such as Malta and Cyprus who, motivated to pursue an independent foreign policy in the 1970s, became members of the non-aligned movement.

Unilateral abstention describes a situation in which one of the superpowers has not involved itself in a region, or at least has refrained from becoming involved in one of the zones of that area. In Southern Europe, the United States involved itself in the core sector, with the Soviet Union concentrating its resources in East Europe and only involving itself in individual peripheral countries, such as Libya for example.

Unilateral abstention often tends to be less stable than mutual non-involvement. This is because, the participation of one superpower in a region usually leads to the involvement of the other as there are few incentives for a superpower to abstain. The only incentives to abstain which may exist are: the fear of stretching one's resources too thinly which could undermine foreign policy objectives elsewhere; the risk of escalating competition to a level that could result in outright conflict; the lack of a convenient inroad into an area.

The third pattern of superpower involvement in the intrusive system is restrained mutual involvement. In this case, both superpowers accept the authority structure (the governing body and its opposition) in power in a particular country or regimes in a certain region and seek to gain influence by working within existing power structures. The parameters that the superpowers operate within are much wider in such circumstances. For example, in the Middle East, the two superpowers settled for a type of "division of labour" formula as they tolerated the acquisition of client states by one another. As time went by, the superpower glue hardened, and a kind of institutionalizing process that reinforced mutual involvement dominated the politics of the region (Macfarlane, 1985: 40).

As indicated earlier, the influence of European powers in the Mediterranean suffered a severe set-back in 1956 when French and British endeavours in the Suez Crisis proved futile. British Prime Minister Anthony Eden had long been waiting for an opportunity to teach Egyptian President Nasser a lesson, in large part due to the latter's anti-western traits (in 1955 Nasser had received armaments from Czechoslovakia, the first Arab nation to do so. In May 1956 Egypt recognized communist China. Nasser's affiliation to the non-aligned movement did nothing to improve his image in the West). In October 1956 an opportunity seemed to present itself. Nasser had nationalized the Suez Canal Company in July 1956, following the American abandonment of his Aswan Dam scheme which seemed to be vital to help combat Egypt's endemic poverty. The British and French responded by calling up twenty thousand reservists for an invasion of Egypt, with Malta and Cyprus designated as jump-off points.

By late October, Britain and France were ready to take action. Though American support could not be assured beforehand, London assumed that Washington would either help when the time came or at least turn a blind eye. But Britain's calculated risk backfired. Just as the Suez stratagem appeared to be tipping in Britain's favour, the United States warned it could exploit sterling's fragile position on the international money markets. Eden was left with little choice. He immediately withdrew British forces from the area and resigned. France withdrew its navy from the isthmus in 1959. Thus the Suez fiasco marked the abatement of the role these secondary powers were to play in future Mediterranean politics. The incident also marks the end of British empire and the rise of US hegemony.

There was little doubt that the writing was also on the wall for Britain's Mediterranean policies when Eden's successor, Harold Macmillan, ordered a defence review upon taking office. The de-colonization process which had really commenced in 1947 with the independence of India, snowballed after a series of speeches by the French and British Heads of State. In an address on 18 September 1959 in Brazzaville, De Gaulle announced he was ready to recognize "self-determination for Algeria" (Times, 19/5/59). Harold Macmillan echoed this theme six months later, during an official trip to Africa. In his speech to the Parliament in Cape Town, South Africa, he concluded on the theme that the "wind of change is blowing through this continent" (Madgwick, Steeds, and Williams, 1982: 286).

Disintegration of the British empire removed one of the most important rationales for a British military presence in the Mediterranean, and hence helps to clarify the relatively rapid rate at which the British disengaged from the region. Secretary of State Iain McLeod shed further light on the reason for the rapid rate of disengagement from the area in a statement in 1964, when he remarked that the British approach to colonial problems underwent a metamorphosis in the latter months of 1959.

The key to this lies partly in Macmillan's belief, confirmed by his election victory earlier that year, that the Conservative Party could only sustain itself in power by developing a modern and progressive profile. Facing topical and profoundly contentious decisions on policies towards Europe and nuclear deterrence, Macmillan was ruthlessly determined to shed the colonial albatross which his opponents were striving to clamp around his neck (Holland, 1985: 227). His African trip in early 1960 was therefore a skilled reconnaissance, prior to a clinical act of disengagement. The granting of independence to Cyprus that year and the blanket decolonization process that followed reflect this change of mood within the Tory ethos.

As Britain gradually disengaged from the area, the United States' Sixth Fleet became the paramount force in the area between 1956 and 1965. In fact, to all intents and purposes, by the mid sixties, they had transformed the basin into an American lake (Kaplan and Clawson (eds.), 1985: 3-177).

By 1965 the process of decolonization in the Mediterranean had come to a conclusion. This resulted in the rise of autonomous political units along the southern shore of the Mediterranean and the emergence of a number of new actors on the scene whose interests and demands now impelled the actions of intrusive, core and peripheral actors. The geopolitical equation in the region became even more complex in the late sixties as the Soviets began to deploy naval forces in the area and also introduced a vigorous outreach programme directed towards the Arabs. The West in general and the United States in particular, deemed this development as a credible threat to Western shipping and oil supplies that either originated or passed through these sea lanes (Luttwak, in Bassioni (ed.), 1975: 39-59).

A succession of events in the 1970s - the oil crisis in 1973, the process of détente, and the suspension of Soviet military expansion in the basin - led the West to completely re-appraise its position in the Mediterranean. This re-evaluation developed a character of its own once it became clear that the European Economic Community (EEC), and Southern European littoral countries were attempting to distance themselves from Washington's policies in the area. This is exemplified by the launching of the EEC's Global Mediterranean Policy in the seventies and the series of bilateral agreements that were signed between European countries and their Arab counterparts.

The Cold War is therefore a period when intensive patron-client relationships developed in the Mediterranean area. Despite the variances in their forms of intervention, the external powers always succeeded in having some degree of influence on its future development. The superpowers could intensify or reduce both co-operative or conflictual relations in the area. Their involvement could promote regional integration or lead to disintegration. In the Mediterranean, the superpower strait-jacket limited regional cohesion and helped spur centrifugal tendencies. The pattern of relations were intergovernmental political, military and economic dominant. The superpower era therefore illustrates the extensive role external actors can play in the international politics of regions.

3.2 Conclusions

There are numerous well-researched chronological and historical accounts of Mediterranean history. For this reason, this chapter does not include any extensive narrative account of them. Instead, the scheme of analysis is applied in a chronological manner with the main aim of providing some insight into how the Mediterranean has evolved as a region (Pryor, 1988).

Such a review of the Mediterranean reveals that general tendencies towards regional transformation have been prevalent throughout. In an attempt to identify the characteristics that influence region-building, this concept has been approached as a unit of analysis in itself. In order to systematically examine this entity, one adopted the three regional modalities outlined in the framework of analysis which enable one to examine the shifts that occur at all levels within the regional system. Thus one is able to monitor any variations in the domestic environment, foreign policy sector and in the structural design. Together these provide an extensive review of the main dynamics at work in region transformation.

The first era discussed was that of classical Greece and the Roman empire. The eighth century saw the beginning of two hundred years of colonial expansion by the Greeks that largely served as a blueprint for the Romans. Even after Greek civilization had been irreparably damaged, its spiritual and intellectual legacy permeated every aspect of Roman life. The Hellenic phase saw the establishment of transnational and intergovernmental relations in the Mediterranean area.

The Romans succeeded in uniting the basin at the height of their conquests, controlling the main trade routes that would guarantee them the necessary resources to ensure their superiority. Their mostly self-contained trading system, providing roads, ports, currency and security, is an example of a high degree of cohesion (social, economic and organizational) and an elaborate network of alliances. Fragmentation of the Roman empire did not result in the complete disintegration of this system of governance in the area.

In the second era examined, Rome's hegemonic grip was replaced by a system that was bipolar in nature, with the Christians commanding the northern shores of the Mediterranean and the Arabs controlling most of the southern coastline. The Mediterranean became a boundary zone and later a conflict based region between two civilizations. Although the balance of power shifted from time to time, particularly when the Arabs registered a series of victories that saw them advance into continental Europe and establish a foothold on the Iberian peninsula, no sole power could repeat the same federal feat of the Romans.

The emergence of the Italian city-states in the eleventh century provided one of the case studies that help to understand the internal dynamics of the peripheral sector. Their erratic foreign policy agenda demonstrates the fluid nature of this classification: at times they approached the core sector (when they co-operated with Rome), and at others they shifted to the periphery (when they aligned with the Arabs). A gradual, but definite devolution of power was hence underway. Much smaller actors now possessed the ability to influence the course of developments. Their main concern nevertheless remained to ensure their own survival in this erratic intergovernmental and transnational type of regional system.

The third era witnessed the expansion of the European nation-states in the Mediterranean which resulted in a multipolar squabble, with Britain emerging as the dominant player. During the seventeenth and eighteenth centuries Britain developed a complex network of intergovernmental relations throughout the Mediterranean. In several ways, this period illustrates the various forms of intrusive sector penetration into regional patterns of interaction which are discussed at length in chapters six and seven.

From an historical perspective, European states did not project power in any significant way before 1500, the exception being medieval Europe's brief Crusade campaign in the eastern Mediterranean during the twelfth century. Like other centres of civilization, Europe was thus predominantly self contained.

British supremacy in the Mediterranean was a unique development in the sense that, for the first time, a non-Mediterranean power supervized the waterways that linked the Atlantic to the Levant. In addition to recognizing early on that the Mediterranean would have to be their main priority if they wanted to remain unchallenged, Britain's success must also be attributed to its ability to delegate tasks to other regional actors. Having skilfully understood the internal dynamics at work in the Mediterranean of the eighteenth century, London established the necessary alliances with the core and peripheral sectors that would ensure its dominant position.

In the section dedicated to developments in the twentieth century, specific attention is given to the features of the intrusive system during the Cold War, i.e., the rich west-north countries against the east and south poorer powers. In addition, this analysis of the antagonism and co-operation inculcated within the Mediterranean by the two superpowers also provides a comprehensive panorama of the pattern of relations operating in the area until 1989. This also serves as essential background information to the final chapter which contends with the process of regional transformation in the post-Cold War period.

After experiencing two World Wars, the Mediterranean again found itself under the scrutiny of a non-Mediterranean power. Initially the United States was reluctant to fill the vacuum being created by Britain's gradual withdrawal from the Mediterranean area. A proactive foreign policy was however adopted, shortly after it became apparent that it would prove costlier in the long-run if they had to return in a crisis situation.

Like interaction within every international region, international politics in the Mediterranean area were eclipsed by the actions of the two superpowers. While the secondary and middle powers of the region (France, Italy, Greece, Israel) formed a weak core that was subordinate to the intrusive system, the smaller littoral states (the peripheral sector) either pledged allegiance to one of the superpowers or else opted for the alternative of non-alignment. Intergovernmental relations therefore developed along geopolitical lines.

Throughout history, concerns of international stability have been paramount in the process of region-building. Stability has meant the preservation of peace and the maintenance of a certain distribution of power. The Mediterranean sea and its surroundings were quite often a battlefield where opposing interests and ambitions clashed. Every power in the regional matrix tried to expand its influence while carefully ensuring that its position in the system was not threatened.

In the classical Mediterranean context, regional variation often consisted of the core (Greece, Rome), dominating its peripheral sector (the entire basin and coastlines), with little or no influence from the intrusive system (Persia). In the medieval Mediterranean context, regional modification witnessed the intrusive system (in particular Britain and later Germany) commanding both the core (France, Italy, the Ottoman Turks) and peripheral sectors (the rest of the basin). Once the region fell into the European continental orbit, Britain employed its superiority at sea to advance further its more general global interests.

In the post-War Mediterranean context, regional transformation again saw the intrusive system (the superpowers) control the international relations of the core (their client states in the littoral) and peripheral (the non-aligned, colonies and minor states) sectors. Thus, once Europe was absorbed into the bipolar Cold War system, the Mediterranean again shared a similar destiny, on this occasion with the United States substituting Britain as the main actor (see Table 3.1).

In the contemporary Mediterranean context, regional change has somewhat ironically acquired the meaning of preserving the status quo. Most of the regional actors are content with the position they currently hold and lack the means and the will to undertake major destabilizing actions in the area. Throughout, mastery at sea has been one of the decisive factors which have contributed to successful campaigns, from the Punic Wars to the Second World War. During the interwar years the Mediterranean was often referred to as a region. The Cold War saw the Mediterranean divided into three regional

PERIOD	DOMINANT PATTERNS OF RELATIONS
HELLENIC	Comprehensive
ROMAN	Federal
CHRISTIAN/MUSLIM	Intergovernmental
ITALIAN CITY-STATES	Transnational/intergovernmental
BRITAIN (EUROPEAN)	Intrusive intergovernmental dominant
COLD WAR	Intrusive intergovernmental dominant
POST-COLD WAR	Intergovernmental

Table 3.1 - Dominant Patterns of Relations in the Mediterranean

blocs, namely Western Europe, Eastern Europe, and the Middle East. In the post-Cold War several authors discuss Mediterranean affairs through subregional prisms, namely Southern Europe, the Levant and the Maghreb (see Table 3.2 for a summary of literary perception of regionality in the Mediterranean area during the twentieth century).

As its name suggests, the Mediterranean is a "sea among lands" and the continental factor has on numerous occasions played a significant role. While a number of the leading players have been maritime powers (Carthage, Venice, Britain, the United States), others have led their expeditions from the land: (Rome, the Arabs, Spain, the Ottoman Turks). This further suggests that no single factor influences the pattern of regional transformation in the Mediterranean.

The Mediterranean evolved from a comprehensive international region in Hellenic times to a federal system under the Romans. The arrival of the Christian/Muslim world resulted in the relegation of the Mediterranean from this category. The Mediterranean initially became a frontier between the North and the South. As patterns of interaction intensified the area developed into a conflict based region. The rise of the Italian city-states saw the revival of transnational patterns of trade extending throughout the area. The dominance of Britain in the seventeenth and eighteenth centuries is a clear example of an intrusive dominant sector in international regional politics. Intrusive intergovernmental patterns of relations superseded contacts littoral states, as first the Europeans, and later the two superpowers, dominated Mediterranean

PERIOD	MEDITERRANEAN AREA	LITERATURE
Inter-War 1919-1939	Mediterranean Region	Silva, Newbigin, Polson, Vannutelli, Petrie, Aymard, Boveri, Hummel & Siewart, Greenwall, Monroe, Ludwig
Cold War 1945-1989	W. Euro-E. Euro-M. East	Bassiouni, Cottrell & Theberge, Kaplan & Clawson, Lewis Jr., Luciani, Luttwak, Pinkele, Pomfret, Rosenthal, Tovias
Post Cold War 1989-	W. & S. Euro-E. Euro-Levant-Maghreb	Aliboni, Buzan, Buzan & Roberson, Agha, Cantori, Miall, Calleja & Wiberg, Fenech

Table 3.2 - Literary Perception of Regionality in the Mediterranean Area During the 20th Century

Source: Fenech, Dominic, *Mediterranean Regionality*, in Fiorini, S., Mallia-Milanes, V. (eds.), *Malta: a Case Study in International Cross-Currents*, Malta University Publications, Malta, 1991

politics. As this trend became a permanent feature of international relations across the Mediterranean, the notion of a Mediterranean international region became more of an historical reality. Comprehensive type relations in Hellenic times developed into centralized exchanges under the Romans. The collapse of the Roman empire saw the fragmentation of these relations along both west-east and later north-south axis (see Table 3.3 for parametric shifts in the Mediterranean).

PERIOD	PARAMETER	DOMINANT CONSTELLATION
Early Mediterranean	Comprehensive International Region	Hellenic
Early Mediterranean	Empire	Pax Romana
600-1517	W/E; N/S	Christian / Muslim
Euro Nation-state	W/E	Spain, Austria / Ottoman
16th. & 17th. century	W/E-N/S	Spain, Austria / France, Ottoman
18th. century	W/E-N/S	Britain, Ottoman / France / Russia
18th. & 19th. century	W/E-N/S	Britain / Russia
Early 20th. century	W/E-N/S	France, Britain / Italy, Germany
Cold War 1945-89	W-N/E/S	USA / USSR
Post-Cold War	(WN)/(ES)	USA, Europe / Russia
21st. century	W/N/E/S	USA (Japan) / Europe / Russia / China?

Table 3.3 - An Historical Perspective of Parametric Change in the Mediterranean

Parametric Definition:

Hellenic: Comprehensive international region
Pax Romana: Federal Empire
Christian / Muslim W/E - N/S: West / East divide; North / South conflict region
Rise of European Nation-state W/E: West / East divide;
16th. & 17th. century W/E - N/S: West / East divide; North / South European divide
18th. century W/E - N/S: West / East divide; North / South European divide
19th. century W/E - N/S: West / East divide; North / South colonial race
Early 20th. century W/E - N/S: West / East divide; North / South colonial empires
Cold War W - N/E/S: West-North / East divide; West-North / South Third World
Post-Cold War (WN)/(ES): West, North synonymous / East, South synonymous
21st. century W/N/E/S: West / North (Europe) / East (China) / South underdeveloped

Whether the Mediterranean area will again revert to a more intergovernmental or even transnational international region type of modality in the post-Cold War world now that superpower overlay has been lifted is the central theme of this investigation. Examining in greater detail developments within and between the Western European and Middle Eastern international regions and their respective subgroupings which border the Mediterranean basin, namely Southern Europe, the Maghreb, and the Levant, will help identify which dynamics are the most cogent in determining the current nature of the regional transformation process.

4 Conceptualizing Regionalism in the Mediterranean Area

The main aim of this chapter is to identify the international regions which operate in the Mediterranean area. This will be carried out by applying the international region definition, outlined in the framework of analysis, to the area in question. An attempt is made to assess the patterns of interaction within each existing international region in an effort to distinguish the nature of the regional dynamics taking place. This is followed by an examination of the distinct subgroupings within each of the international regions which border the Mediterranean, namely Southern Europe, the Levant, and the Maghreb (see chapter 2, section 2.2). After examining the patterns of interaction within each of these units, this section concludes with a review of the connections and disconnections between the three Mediterranean hinterlands in an effort to detect regional dynamics which would suggest the re-emergence of a Mediterranean international region.

The following criteria were identified in the framework of analysis as necessary for an area to qualify as an international region: the states' pattern of co-operative or conflictual relations or interactions exhibit a particular degree of regularity and intensity to the extent that a change in their foreign policy actions have a direct influence on the policy-making of neighbouring states; the states are proximate; the region consists of at least two but probably more states; the influence of intrusive action is considered.

When this set of criteria is applied to the Mediterranean area, two prominent groupings of states emerge: namely the geographical space which borders the north-west sector of the Mediterranean which is labelled Western Europe, and the geographical area covering the south-eastern flank of the basin and is labelled the Middle East.

4.1 The Western European International Region

The term Western Europe, as used in this study, does not coincide with the geographic definition: it includes all the European Union member states, plus Turkey, Malta and Cyprus. Although geographically Turkey lies on the periphery of the European continent, it is included in the European international region because the secular choice of Kemalist nationalism has brought the country into the Western and European sphere. Institutionally, Turkey is a decisive component of NATO as demonstrated during the Gulf War in 1990-91 and it is closely linked to the European Union, particularly after negotiating a customs union with Brussels in the mid-nineties (Aliboni, 1992: 1-2; Buzan, 1991a: 196). In short, Turkey shares strong intergovernmental and transnational links with Western Europe.

The inclusion of Malta and Cyprus is based on the two Islands' historical, cultural, economic and political ties with Europe. Both countries applied to become full EU members in 1990 and their applications are currently being processed by the European Commission (ibid.).

Thus, the concept of a Western European international region adopted in this thesis is geopolitical. As members or prospective members of the EU, all of the countries within this catchment area share similar domestic and foreign policy concerns and have pledged to integrate their policies in accordance with the regulations of the Maastricht Treaty. This international region also shares a similar line of ethno-cultural thinking, with an emphasis on the role of Latin Christendom in defining a community of states (Koenigsberger, 1987: 136- 210, and Pirenne, 1936: 1-46). The area also meets the other international region criteria in that it consists of more than two states, the countries considered are generally proximate in geopolitical terms, and all of the countries are affected by similar external constraints (they are members of the Western security system). Above all else, this international region is distinct from the areas in its vicinity because of the particular intensity of intergovernmental and cross-border interactions between the countries listed (Miall, 1994: 1-3).

Objections can always be raised regarding the cast of actors included in an international region (Buzan, 1991a: 196-200). In such an exercise, the major methodological problem remains in drawing boundaries between such regions. One way to overcome the complexity of boundary delineation and simultaneously test the validity of the proposed international region is to verify that the intimacy of interaction among the participating states begins to wane as the 'edge of one international region and the start of the next is approached'. This test certainly reinforces the notion of a Western European international region, as this area shares similar constellations of political and economic patterns which contrast sharply with the patterns of interaction found in the adjoining international regions of Central and Eastern Europe and the Middle East. It is the latter international region which mainly concerns us in this analysis of the Mediterranean area. Although Central and Eastern European patterns of relations influence regional dynamics in the Mediterranean, particularly those which involve Albania, Croatia and Bosnia, they are excluded from this analysis because events in this area have a more direct impact on European relations than those of the Mediterranean. As a result, their inclusion will not add any new dimension to the already comprehensive list of countries being assessed in this research.

During the Cold War, the Western European international region was an area of uneven development. The creation of the European Economic Community in 1957, and the rival European Free Trade Association shortly afterwards, symbolize the alternative visions shared by European policymakers in the post war period. The enactment of the Single European Act in 1986 and ratification of the Maastricht Treaty in 1993, coupled with the various phases of EC enlargement (1973, 1981, 1986, 1995) have combined to make Western Europe, i.e. the EU and associate members, an intricate intergovernmental and transnational international region. The degree of political and economic integration achieved is evident when one notes that accepting to abide by the EU's *acquis communitaires* is equivalent to agreeing to 14,000 pages of EU legislation (Calleya, 1993: 37).

Interdependence among the states of Western Europe is not a new phenomenon. Intergovernmental ties are exemplified in the multiplicity of political, economic and security institutions, such as NATO, the OSCE, the EU, and the WEU. What has changed is the extension of globalization since the 1970s which has led to a rapid development of the networks of links between these states (Miall, 1994: 5; see Graph 2 which denotes regional percentages of world trade). Transnational patterns of interaction have become a fundamental fact which characterize the political and economic structure of the world, at the end of the twentieth century.

Western Europe is still organized politically into nation-states with sovereign governments, but increasing integration between these nations is gradually eroding the differences among national economies and undermining the autonomy of national governments. Profound technological, social, and cultural changes have brought Western European states closer together by reducing the effective economic distance among them (Bryant, 1994: 42).

Exchanges take the form of trade, investment, capital flows, cross-national corporations, large-scale movements of people, dense patterns of rail, road, sea and air traffic, and an instantaneous sharing of information, news and media (Wallace, 1990: 8-12). The European international region is specifically characterized by these intricate transnational patterns of interaction. When coupled with already existing intergovernmental ties, this group of countries clearly demonstrates the attributes of a quasi-comprehensive international region, sharing similar social movements, economic shocks and political developments.

Graph 2

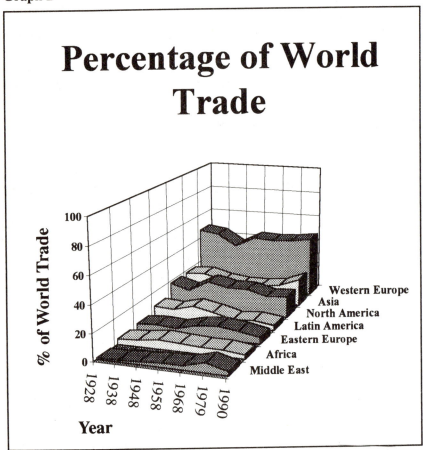

Source: Anderson, K. and Blackhurst, R. (eds.), *Regional Integration and the Global Trading Systems*, Harvester Wheatsheaf, 1993: p.29

Immediate developments in Western Europe during the post-Cold War period have been the attempts to consolidate economic and security co-operation. By 1996, convergence in both of these sectors is still a distant goal. However, the creation of a single European currency and the establishment of a common foreign and security policy remain high on the political agenda and will be discussed at length at the EU intergovernmental conference scheduled for 1996/97. In the interim, it appears likely that a core group of European Union countries could introduce a single currency. In security, Western European states are increasingly showing a willingness to co-ordinate their foreign policies. This is reflected in the similar priorities this group of countries attaches to external relations: bilateral relations within the Western European international region are most important; followed by European Union affairs which are linked to EU ties with Eastern Europe, North America and Asia; ties with the Middle East are a distant third and it is within this category that Mediterranean issues are considered (see Table 4.1).

4.2 The Middle Eastern International Region

At first it might appear misleading to talk about the Middle East as an international region, especially if one mistakenly perceives the Middle East as being on a par with Europe in being defined as an international region. Yet, as outlined in the framework of analysis, international regions vary in both nature and degree. Thus while the European model conforms to a certain modality of regionalism (co-operative transnational, intergovernmental and comprehensive dominant) the Middle East entity is of a completely different nature (conflictual intergovernmental and transnational dominant).

The term Middle East, as used in this study, does not coincide with the geographical definition. It stretches from Morocco to the Persian Gulf. It is distinct from South Asia which lies to its east and black Africa which lies south of the Sahara (ibid.: 196). The Middle East international region includes the following subgroupings: the Maghreb, which is referred to usually as North Africa in the West but which is considered as part of the collective consciousness of the Arab Middle East (especially in an historical sense), the Levant, also referred to as the Mashrek, or the Fertile Crescent, which in this study includes Egypt - and the Arabian Peninsula including the Persian Gulf which incorporates Iran.

SUBGROUP	Geo-Strategic Priorities			
	1st.	2nd.	3rd.	4th.
Southern Europe	Western European Bilaterals	European Union (Eastern Europe) (USA) (ASIA)	Middle East Bilaterals	Mediterranean Issues
Maghreb	Maghreb Bilaterals AMU and Arab League	European Union	North America (USA) ASIA (Japan)	Mediterranean Issues
Levant	Middle East Bilaterals Arab League	USA	European Union	Mediterranean Issues

Table 4.1 - Mediterranean Subgrouping Foreign Policy Priorities

Although geographically Egypt is located along the North African coastline, it is included in the Levant because of its consistent northern policy concentration dating back to the nineteenth century and its clashes with the Ottoman Empire (Selim, 1994: 2). Throughout the latter part of the twentieth century Egypt has also played a key role in intergovernmental relations across the Levant, involved in all of the major wars in the area and setting a precedent in the Arab World when it recognized Israel in the Camp David peace talks (Ajami, 1987: 89-155).

The one salient reality that most unites this area, is the Islamic religion and the history of the Islamic Empire. This common denominator has more recently been supplemented by what may be described as the transnational political force of Islam (Buzan, 1991: 196). This phenomenon, perhaps more than anything else, combines to create a potent Middle East international region. Political Islam effectively challenges secular European nationalism which dominates an area that was Islamic for well over a millennium (see Appendix for map denoting the spread of Islam). One objective shared by Islamic groups throughout this area is a return to a societal system that is indigenously Islamic. Seen from such a perspective, a secular national state is a continuation of external rule by other than direct physical means. It is at this political level, rather than at a religious one, that a clash between the West and Islam lies (Agha, 1994: 241-2).

As a people, Arabs enjoy a high degree of linguistic and cultural homogeneity. Despite the distinct regional dynamics at play within the various Middle Eastern subgroupings, most Arabs belong to the same religion, converse in the same language, and share the same cultural traditions (Korany and Dessouki, 1991: 33). These facts gave rise to the pan-Arab nationalist trend which dominated intergovernmental regional proceedings throughout the 1950s and 1960s (Vatikiotis, 1984: 77-116). They also played a significant role in the anticolonial movement in Africa and in the development of the non-aligned movement.

Like its Western European counterpart, the concept of a Middle Eastern international region is geopolitical. However, the patterns of interaction between Middle East actors are much more erratic and more conflictual than co-operative in nature (see Table 4.2). This is not to deny that they may occasionally share certain concerns - primarily related to security - or might co-operate in specific fields, as has been the case in the energy sector. But the track record of the Arab League and other efforts to institutionalize relations in the Middle East, such as the Gulf Co-operation Council (GCC) and the Arab Maghreb Union (UMA), clearly demonstrate the limits of co-operation among the majority of the countries within this international region (Salame, 1988: 256-79).

INTERNATIONAL REGION	MEMBERS		DYNAMICS	
Western Europe	EU States	Malta	Comprehensive	⇑
	Turkey	Cyprus	Intergovernmental	⇑
			Transnational	⇑
Middle East	Maghreb	Levant	Comprehensive	**X**
	Persian Gulf	Arabian Peninsula	Intergovernmental	⇑
			Transnational	⇑

Code:
⇑   Active
⇔   Neutral
⇓   Decline
X   Non-existent

Table 4.2 - International Regions Encompassing the Mediterranean Area

Although each Middle East subgrouping has its own node of security interdependence with their own distinctive dynamics, there is enough cross-border interaction within the Middle East to justify identifying it as a regional unit of analysis. Even though opposition towards Israel has somewhat waned, the two dozen states in this region continue to share similar political and socio-economic challenges. These grievances are often voiced at Arab League meetings which provide a legitimizing forum in which the affairs of the different Middle Eastern subgroupings are brought together.

The Middle East also meets the other criteria associated with an international region. It consists of more than two states and the countries considered are generally proximate in geopolitical terms. In sharing similar economic, political and cultural patterns, the Middle East is also affected by external challenges which emanate from Western secularism. As emphasized above, it is the intensity of intergovernmental interactions between the countries in this area which qualifies the Middle East as an international region. When compared with the degree and nature of relations which exists across the Mediterranean, the distinct attributes of this international region become even more pronounced.

A comparative analysis of political and economic patterns in Western Europe and the Middle East reveals that comparable patterns of interaction are conspicuously absent (see Graph 3 for details of intra-regional trade patterns). Western Europe and the Middle East developed very different patterns during the Cold War. Intergovernmental ties in the Middle East were relatively strong as was reflected in the creation of the Arab League in 1944-45, which quickly grew to include twenty-one states (Buzan and Roberson, 1993: 136). The process of decolonization and independence facilitated pan-Arab nationalism and Islam's ability to create transnational political linkages across the whole international region. In the period between achieving independence and the end of the Cold War, the Middle East consolidated itself further as a distinct and mainly self-contained international region (ibid.: 137). Most of the states have been too preoccupied with distinct domestic or regional subgrouping security dynamics to attempt nurturing a complex network of relations with all the states in the region.

Since the end of the Cold War, older patterns of relations have been released, as manifested in the force of political Islam. There has also been a shift in the patterns of amity and enmity within the region. Relations in the Levant appear to have momentarily improved following breakthroughs in the Arab-Israeli peace talks. In contrast, intergovernmental ties in the Maghreb are at an all time low, with the crisis in Algeria and the Libyan-Lockerbie affair becoming fixed diplomatic impasses. A review of post-Cold War events thus reveals that there is no consistent pattern of relations in the Middle East. In the Levant, political developments since the end of the Cold War have tended to increase the level of interdependence. In the Maghreb, the opposite process has taken place.

Transnational flows in the Middle East are much less dense than Western Europe and are more controlled. North-South non-state economic links radiate around the energy sector. Pipelines across the Maghreb stretch across the Mediterranean to Europe. Similar projects link the Persian Gulf to Europe and the Levant. But no intricate pattern of transnational interdependence involving trading links, cross border investment and common institutions has yet been developed. South-south transnational patterns of interaction are mainly limited to the area of Islam. This low level of interdependence partly explains the high level of enmity in Middle Eastern patterns of relations. Interdependence does not in itself determine either co-operation or conflict, but it does increase the stakes in relationships. The more a sense of common interests is nurtured and the benefits of participation in a larger community are apparent, the less likely the outbreak of hostilities. This in fact is one of the reasons why the European Coal and Steel Community (ECSC) was set up in 1952.

A key difference between Western Europe and the Middle East is that,

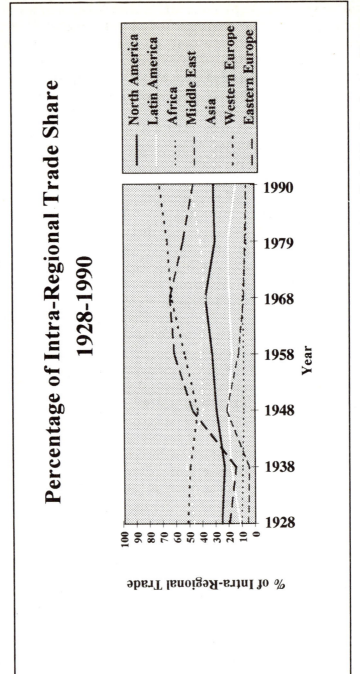

Graph 3

Percentage of Intra-Regional Trade Share 1928-1990

Source: Anderson, K. and Anderson, R. (eds.), *Regional Integration and the Global Trading System*, Harvester Wheatsheaf, 1993: p.29

whereas the former has developed a multi-level international society in which international institutions, states and sub-national organizations all play important roles in managing cross-national transactions, the latter has not. Within the Middle East, the level of cross-national transactions is much lower and has not been institutionalized in any comprehensive manner. For example, the Arab League met with success in only six of seventy-seven conflictual situations it attempted to settle between 1945 and 1981 (Awad, 1994: 153). Similarly, other co-operative initiatives undertaken by subregional organizations, such as the Arab Maghreb Union, have never achieved any significant degree of real political or economic consensus. One can also say that centrifugal forces have superseded centripetal forces as the majority of states in the Maghreb see their future in securing market access to Western Europe.

In the Middle East, post-Cold War efforts to increase economic and security co-operation are still at an embryonic stage. The international Middle East economic summits at Casablanca in October 1994 and Amman in October 1995 displayed many of the obstacles that would have to be overcome before this international region could begin to develop stronger transnational and institutional links (Blair, 1995: 4-5). No measures have been introduced to tackle the indebtedness of certain Arab countries, which amounted to $142 billion in 1989 with a yearly debt service of $14 billion, or 33 per cent of the export earnings of these countries (Hitti, 1994: 89). In addition, within the Arab world the gap between the "north", which groups the nine oil-producing countries with small populations (except for Algeria and Iraq), and the "south", which groups the remaining twelve Arab countries, remains wide. The gross domestic product of the former group reached $300 billion in 1990 compared with a GDP of only $119 billion in the latter group (ibid.: 90). A shift to more co-operative indigenous political practices must therefore be coupled with large influxes of foreign direct investment if the necessary infrastructural changes are to occur so that transnational links can be strengthened. Perhaps valuable lessons can be learned from the 1970s, when economic links within the Arab world were considerable as a result of the boom in the petroleum industry. Intra-regional financial flows floundered in the 1980s as a decline in the demand for oil and the consequent slump in revenues settled in (Noble, 1991: 58).

In the Middle Eastern international region in general, there is still no sign of convergence in foreign policy priorities. Each regional subgrouping continues to dedicate most of its diplomatic resources to local security challenges. For example, Syrian relations remain focused on developments within the Levant, while Moroccan contacts continue to focus on bilateral relations in the Maghreb. Contacts with the European Union continue to gain in prominence, although the Levant still concentrates on maintaining strong ties with the United States. Like Western Europe, Mediterranean issues are a lower priority in the chain of foreign policy priorities (see Table 4.1).

These contrasts highlight the differences in political, socio-economic and historical development between Western Europe and the Middle East. While some societies and states across the Mediterranean are in the West European sphere of influence, others find themselves in the Middle Eastern one. The Mediterranean acts as a kind of divide between the countries in the area, several of which are at different stages and perhaps on different paths of historical evolution. In the short to medium term there seems little prospect of Middle Eastern societies rapidly adjusting to West European patterns or vice-versa. Does this mean that it is impossible to develop a common Mediterranean order, that may serve as a precursor to a Mediterranean international region? Or is the developing pattern of transnational, intergovernmental, and institutional links at least capable of creating enough co-operation to limit conflicts across the Mediterranean divide? In the absence of any strong movement towards regionalism among the economies of the Western Europe and the Middle East international regions, it seems sensible to examine the patterns of relations within the context of the three subregions bordering the Mediterranean, that is, Southern Europe, the Levant and the Maghreb, and then attempt to identify any trends towards regionalism in the area.

4.3 International Regional Subgroupings in the Mediterranean Area

International power relations are in constant flux; older power centres decline and new ones emerge for reasons that we only partially understand. The laws of economic motion appear to be one of the important factors in determining the processes of decay and revitalization. A central feature in the transformation of the present international system is the stabilization of some regional power centres and the emergence of new ones. This process is not actually new: it is an aspect of the general dynamics of international relations in which the relative strength of certain powers declines and new ones gradually emerge. A distinct feature of the current transformation is the fact that regional powers are now emerging outside the traditional core of the international system.

By the end of the nineteenth century, European dominance was so absolute that it not only created a single global international system, but it also destroyed older regional patterns of interaction. Regional dynamics were largely absent from the international scene for most of recent history, as the Europeans pursued their own regional dynamics on a global scale. During the twentieth century, this extreme pattern of systemic dominance began to unravel (Buzan et al., 1994: 7-8).

The period 1914 to 1989 saw the self-destruction of European, and to a certain extent Western, world power. The decolonization process between 1945 and 1965 set free the majority of remaining colonies in Asia and Africa. This permitted the re-emergence of regional patterns of interaction, albeit in a significantly different systemic context from that which had prevailed between 500 BC and 1500 AD. In the ancient and classical world, international regions were the dominant structure in the international system. The period of Western domination saw the system level almost eliminate international regional patterns of interaction altogether. During the latter half of the twentieth century, the decolonization process coupled with the collapse of the Cold War created conditions conducive to a resurgence of regional dynamics in the system.

In the Mediterranean area, the decolonization process released many older patterns of regional international relations, somewhat veiled in new Western clothes. Countries such as Egypt and Morocco had roots that stretched back into ancient and classical times. Independence ushered in a period when old rivalries and fears could again be addressed. Apart from being directly influenced by the withdrawal of colonial powers from the Third World, the evolution of regional centres was also spurred by attempts at regional integration among developing countries and the development of economic and political forces within these new states. In addition, newly independent states emerged into an environment that was mainly under superpower and Western control. A large number of the new states therefore remained tied to some degree to the ex-metropolitan powers. In the Mediterranean, countries such as Tunisia, Algeria and Morocco remained dependent on France for both their currency and their military security (ibid.).

Around the Mediterranean basin, distinct regional dynamics also developed during this period. Along the northern shore, Southern European states became more intertwined in relations with their northern neighbours through the process of European integration. The admittance of Greece into the EC in 1981 and of Spain and Portugal in 1986, shows how this area's pattern of interaction became European dominant.

SUBGROUPINGS	MEMBERS		DYNAMICS
Southern Europe	France	Greece	Comprehensive ⇑
	Spain	Turkey	Intergovernmental ⇑
	Italy	Cyprus	Transnational ⇑
	Portugal	Malta	
Maghreb	Morocco		Comprehensive **X**
	Algeria		Intergovernmental ⇔
	Tunisia		Transnational ⇑
	Libya		
Levant	Egypt		Comprehensive **X**
	Israel		Intergovernmental ⇑
	Lebanon		Transnational ⇑
	Jordan		
	Syria		
	Palestinian entity		

Code:

⇑   Active

⇔   Neutral

⇓   Decline

X   Non-existent

**Table 4.3 - International Regional Subgroupings in the   Mediterranean
    Area**

In contrast, no such firm patterns of regional interaction occurred along
the southern and eastern shores of the Mediterranean. Most of the new states
in these areas were too weak to generate a durable pattern of regional relations
in their first years as independent states. The majority of them were more
concerned with domestic infrastructural problems and dedicated whatever
resources they had, to ameliorating their internal situation.

As outlined at the start of this chapter, the picture that emerges around, but
not across the Mediterranean, at the end of the Cold War, is of two international
regions: a quasi-comprehensive Western European international region and
an intergovernmental dominant Middle East international region with limited
transnational ties. Both entities incorporate subgroupings. As stated in chapter
two, the term "subgrouping" refers to states within an international region
which share exclusive commonalties with one another, i.e., subregions, and
subsets of international regions which do not necessarily share enough

commonalties to qualify as a subregion. The subgroupings which surround the Mediterranean are Southern Europe (subset of Western Europe), the Levant and the Maghreb (subregions of the Middle East). Comprehending the degree and types of interaction within these subgroupings is important because they are the primary points of contact between the international regions which encircle the Mediterranean (see Table 4.3).

4.3.1 Degree and Types of Interaction in Southern Europe

The Southern European label as used in this section does not coincide with the geographic definition. The concept of a Southern Europe area is geopolitical and is based on two main criteria: all the countries belong or are associated with NATO or the European Union and all states referred to border on the Mediterranean. This subregion thus includes, from west to east, Portugal, Spain, France, Italy, Malta, Cyprus, Greece and Turkey.

Although Southern European countries, particularly those who are already members of the EU, are perceived as sharing common political and economic objectives with other non-member Mediterranean states, this perception of solidarity is not strong enough to be reflected in permanent political structures (Aliboni, 1992: 2). Throughout the Cold War, the large number of actors with diverse political agendas and the geographic discontinuity of the area added to the sense of political fragmentation.

The end of the Cold War has assisted in removing some of the first factor. The emergence of Central and Eastern European countries from behind the Iron Curtain has resulted in the polarization of the Southern European grouping who are now in direct competition for foreign direct investment and assistance from the EU. This has also generated perceptions among non-EU Southern European states, such as Malta and Cyprus, of isolation and marginalization. Members of the EU's southern flank have reacted by starting to co-ordinate their political agenda much more in an effort to prevent Brussels from shifting its external policies lenses totally to the east. The three consecutive EU presidencies of France, Spain, and Italy during 1995 and 1996 demonstrate the significance these Southern European countries attach to the Mediterranean area. Their concerted approach to crisis situations in the littoral, such as that of Greece, Turkey and Cyprus, and the civil war in Algeria, reveals the common foreign policy concerns that each country shares.

A second unifying factor is that the end of the Cold War has weakened the role of the non-aligned movement in this area. Countries such as Malta and Cyprus are currently on the EU 'waiting list', having submitted their applications to join the EU in 1990. Foreign policy patterns in the Southern European area therefore reflect a tendency to becoming more homogeneous, i.e. members of similar political and military organizations: NATO-EU-WEU-OSCE. As the weighting of Mediterranean members increases in these security arrangements, trans-Mediterranean policies could receive more attention in European security debates and simultaneously assist in moving Mediterranean issues further up the Western European foreign policy priority chain (see Table 4.1).

A review of bilateral interactions within this subgrouping now follows in an attempt to identify what dimension they contribute to regional dynamics across the Mediterranean. This assessment will be followed by a similar analysis of the networks of bilateral relations in the Levant and the Maghreb.

France

The state that wields the most power in the Southern European subgrouping is France. In fact, France remains a continental European, an Atlantic, and a Mediterranean power. Although the Mediterranean dimension of French security policy is often limited, its involvement in any Southern European scheme is essential if a proposal is to have any chance of success. This is evident if one examines the success rate of the various trans-Mediterranean security initiatives launched since the end of the Cold War. While the CSCM remains a distant aspiration, the French Western Mediterranean Forum got off to a rapid start and stalled after France's reluctance to turn a blind eye on allegations linking Libya to acts of terrorism.

The main reason for France's dominance in the Southern European subgrouping is its superiority in economic, political, and military terms. France's position in the core sector of this group is bolstered by its nuclear deterrent force, its levels of military expenditure, and its power projection capacities. Moreover, France's current relations with the Maghreb countries, especially Algeria, its colonial past and the significance of North African immigrants in the French community all make Mediterranean and Southern European issues an integral part of the French national identity (Carle, 1992: 41).(A review of France's role as an internal great power in the Mediterranean is provided in chapter six.)

Spain and Italy

Since the fall of the Berlin Wall, France's position in Southern Europe has occasionally been challenged by Spain and Italy. Both have shown an increasing interest in becoming more active in creating stronger links with their counterparts on the southern shores of the Mediterranean. But a string of largely domestic episodes have prevented both Madrid and Rome from following through and realizing their external agenda. As a result, Spain and Italy remain in the semi-peripheral antechamber.

Spain's primary concern remains over the crisis in the Maghreb. It has made strenuous efforts to enhance its reputation in North Africa following the Gulf War. For example, it has implemented a series of bilateral agreements, including an award of $1 billion worth of export credits to Algeria in 1992-93 (Blunden, 1994: 140). In February 1996 Spain and Morocco agreed that they would commence with the first phase of a 22km tunnel under the Straits of Gibraltar in 1997 that will facilitate bilateral trade between the two countries which currently exceeds $1 billion.

Spain has also argued consistently that the North-South dimension of security challenges have not been addressed by existing international security institutions. It is for this reason that it joined Italy in 1990 in announcing the necessity of creating a regional framework for dialogue and co-operation in the Mediterranean, namely the Conference on Security and Cooperation in the Mediterranean (CSCM).

Spain's geographic position, and the political and economic capital it has managed to achieve as a member of the EU, dictates that this Iberian country will continue to play a key role in Southern European affairs (see Agarwal et al., 1994: 346-58, and Table 4.4, World Bank Statistics, 1977, 1986, 1992; and Table 4.5, *The Military Balance, 1993-1994*). Militarily, Spain is entrusted with ensuring access to and the safety of passage through the Straits of Gibraltar. It also has the resources to conduct tactical naval and air operations in the Western Mediterranean. Spanish military bases are also useful transit points for operations conducted in the eastern Atlantic or eastern sector of the Mediterranean (Rodrigo, 1992: 103-4).

Prior to the domestic political revolution which has engulfed Italy in the nineties, Rome's defence policy was also focusing more and more on the security dimension of North-South relations. In 1980, Rome signalled its intention to develop an active foreign policy in the Mediterranean when it decided to guarantee Malta's neutrality. At the end of the eighties Italy promoted the idea of a stable co-operation framework stretching from the

SUBGROUP	REAL GDP ($Bn)			RELATIVE CHANGES IN REAL GDP	
	1977	1986	1992	1977-1986	1986-1992
MAGHREB	43.31	116.23	113.30	168%	-3%
Mauritania	0.21	0.70	1.20	232%	72%
Morocco	9.50	16.07	28.34	69%	76%
Algeria	10.10	70.00	38.70	593%	-45%
Tunisia	5.00	8.84	15.82	77%	79%
Libya	18.50	20.62	29.24	11%	42%
LEVANT	36.90	116.44	124.18	216%	7%
Egypt	13.30	62.00	40.89	366%	-34%
Israel	14.20	29.38	64.67	107%	120%
Lebanon	2.90		2.82		
Syria	6.50	25.06	15.80	286%	-37%
S. EUROPE	783.65	1,688.03	3,359.35	115%	99%
Turkey	46.60	58.07	118.90	25%	105%
Greece	26.30	43.17	76.37	64%	77%
Italy	193.70	599.92	1,222.40	210%	104%
France	374.80	724.62	1,270.50	93%	75%
Spain	123.60	229.10	578.36	85%	152%
Portugal	16.40	28.99	83.47	77%	188%
Malta	0.75	1.30	2.67	73%	105%
Cyprus	1.50	2.86	6.68	91%	134%

Table 4.4 - GDP Statistics in the Maghreb, Levant and S. Europe (World Bank, 1993)

shores of the Mediterranean to Central Europe when it joined the "Esagonale" (Hexagonal Group) with Austria, Yugoslavia, Hungary, Czechoslovakia, and Poland (Greco and Guazzone, 1992: 73). After the Gulf War, Italy championed the notion that the time had come to fill the security vacuum in the Mediterranean basin. The concept of a CSCM was presented as one possible way of addressing the endemic political and economic instability in this area.

This concept of comprehensive security is the leit motif of all of Italy's Mediterranean policies. Integrating economic, political and military means in a global co-operative strategy towards the Mediterranean is regarded as a much more effective approach than ad hoc bilateral actions. It is for this reason that Italy continues to prefer the EU as the main locus of European security and is reluctant to see an extension of NATO responsibilities in the Mediterranean (ibid.: 78). Italy and France are also keen to develop a federalist approach to EU decision-making in security policy, with the WEU becoming the main security institution in the basin. Other Southern European countries, such as Spain, prefer a more intergovernmental approach in this sector (Aliboni, 1992: 13).

The melting of the Cold War glacier has therefore once again allowed the Southern European core and semi-peripheral countries to play a much more direct role in the Mediterranean. After all, all three countries were at the forefront of the "scramble for Africa" at the turn of the century when the Europeans divided North Africa amongst themselves. Their participation in the Western European Union is enabling these countries to institutionalize their naval co-operation in the area by formulating contingency combat plans. The three Southern European Union member states and Portugal established a joint land force, known as Eurofor, with a headquarters in Florence, and a naval force called Euromarfor, led by a French aircraft carrier at the start of 1995. The new forces are to be employed within the WEU framework, but are also at the disposal of NATO (Financial Times, 2-6-95: 2; see also Financial Times, 27/28-11-93: 2).

The setting up of a Medcorps between France, Italy, Spain and Portugal partly offsets the predominant position of the United States' Sixth Fleet as each of these European states operate naval aircraft carriers (*WEU Report*, 15-6-94). The launching of such a southern security structure assists the Southern European governments assert a Southern Mediterranean orientation in the EU to counter what they perceive as an increasingly Germanic Europe, with its centre of gravity steadily moving north and east after the Nordic enlargement.

The core and semi-peripheral sector of the Southern European

ARMED FORCES	Mauritania	Morocco	Algeria	Tunisia	Libya	Egypt
ARMY						
Total Manpower	15,000	175,000	105,000	27,000	40,000	310,000
Infantry Brigades		3	5	3	3	8
Armoured Personnel Carriers (SAM, SSM)		785	460	268	1,040	3,575
Artillery Pieces (Towed)						
Armoured Brigades	1		2	3	11	4
Tanks	35	284	960	84	2,300	3,167
Paratroop Brigades	1	2				1
AIRFORCE						
Total Manpower	150	13,500	10,000	3,500	22,000	30,000
Bombers						
Fighters			129			
Transports	2		21	2		25
Helicopters		24	7	5		74
Combat Ready Aircraft	7	93	193	32		546
NAVY						
Total Manpower	400	7,000	6,700	5,000	8,000	20,000
Destroyers						1
Submarines			2		5	2
Torpedo Boats				6		
Patrol and Escort	6	23	8	14	39	18
Landing Craft		3			5	
Minesweepers			1		8	8

Table 4.5 - Balance of Forces in the Maghreb, Levant and S. Europe

Israel	Lebanon	Syria	Turkey	Greece	Italy	France	Spain	Portugal
134,000	40,000	300,000	370,000	113,000	223,300	241,400	138,900	27,200
9	1			1				3
6,000		1,500	2,896	2,165	3,683	4,100	2,000	357
3					1		1	
3,960	350	4,500	4,835	2,640	1,210	1,000	1,148	209
1						1		
32,000	800	40,000	60,000	26,800	77,700	90,600	29,800	11,000
	3							
59	2	29						
93	5	100						
662	3	639	539	384	385	796	150	99
10,000	500	8,000	50,000	19,500	43,600	65,400	32,000	12,500
			11	8	4	4		
3		3	15	10	8	19	8	3
		21	16	10	6			
44	5	11	29	40	18	23	28	16
		3	7	12	2	9	4	3
		9	32	14	12	21	12	

(*The Military Balance*, IISS, 1995)

subgrouping see their dominant positions as dependent on the stability of North Africa. "Stability" in this sense means the avoidance of large-scale unrest and preservation of the status quo. Tensions between the Southern European and the Maghreb regional subgroupings are largely caused by economic, political, and cultural, rather than military issues. For example, a full-scale civil war in Algeria could generate a flood of refugees across the Mediterranean that would destabilize internal security in countries such as France. Another threat, that Europe has to make contingency plans for, is a territorial clash - for example campaigns concerning the Spanish enclaves of Ceuta and Mellila in Morocco and the Canary Islands could quickly intensify. Economic collapse is another substantial threat that has to be considered - the protection of important export markets and the security of large outstanding debts is vital if European economies are to continue their gradual but fragile recovery from recession. For example, in 1992, Algeria's external debt was approximately $25 billion, of which more than $6 billion was owed to France (Swan, 1992: 7). The fact that the major creditor, Coface - the French export credit department, had to write off its loans to Iraq after the Gulf War makes the protection of its investments in North Africa all the more compelling.

The Spanish and Italian CSCM proposal and the participation of all three states in the Western Mediterranean Forum ("5+5" talks), coupled with the increasing number of joint manoeuvres between the naval and air forces of France, Italy and Spain, demonstrates the active interest of these Southern European states in contemporary international relations of their southern periphery. The strongest link between the Maghreb and Southern Europe is to be found in the energy sector. A gas pipeline is currently being constructed connecting Algerian gasfields, through Morocco, with Spain and eventually France. Another pipeline linking Algerian gasfields, with Italy, via Tunisia and the Straits of Sicily was opened in 1983 and recently underwent adjustments to double its throughput capacity. These examples make it clear that the energy sector is the one area where co-operation across the Euro-Maghreb space is most active. Imitation of such co-operation in other areas, such as trade, investment, and technological exchange, would enhance economic solidarity between the two Mediterranean subgroupings and help to intensify co-operative north-south transnational relations (see Abdelkader Sid Ahmed, in Marie-Lucy Dumas (ed.) 1992: 154).

Greece

Other candidates that sometimes play roles that qualify them for inclusion in the Southern European semi-periphery are Greece, Turkey and Portugal. Their level of power permits them to execute foreign policy options that are primarily related to their immediate spheres of influence. Unlike the core and other more active semi-peripheral states in Southern Europe, their approach to external affairs is much more erratic and limited. Greece's introduction of an embargo against FYROM (also called Macedonia) at the time of its EU Presidency in 1994 is indicative of such inconsistent patterns of external relations adopted by countries in this sector. Athens's threat of vetoing the EU customs union with Turkey is another example of such an erratic approach (Financial Times, 10-2-95: 2).

In addition, Greece has not been able to use its EU membership to improve its economic position vis-à-vis the other EU member states (see Agarwal et al., 1994: 346-58; and Tables 4.4 and 4.5). This may be partly explained by the fact that Greece spent nearly $2 billion on weapons in 1992, ranking it as the second largest arms importer in the industrialized world for the period 1988-1992 (Hockenos, 1994: 22). On a different note, Athens has somewhat succeeded in diluting its "bad boy" image by forcing Mediterranean security issues back on to the European Union agenda, when it held the EU presidency in 1994 (Agence Europe, 26-6-94: 1 and 4). At the Corfu Summit, EU members endorsed the idea of convening a European-Mediterranean conference, which was convened during the Spanish EU presidency in November 1995 (ibid.).

Turkey

As already noted, Turkey is geographically and politically in the peripheral sector of the Western European international region, and the more specific Southern European subgrouping. It occupies an insulating position between the adjoining European and Middle Eastern regions (Buzan, 1991a: 196). As an Islamic, but not Arab, secular state, and as the former imperial power in the Middle East, Turkey has no desire to become completely engulfed in the Middle East pattern of relations (Buzan and Roberson, 1993: 136). The conflict situations and underdeveloped neighbours to its south and east provides Turkey with plenty of incentive to pursue further its contacts with the West European international region.

As a full member of NATO and the OSCE and an associate member of the EU, Turkey already has strong intergovernmental links with Western Europe and the United States (Hockenos, 1994: 21-2). The challenge confronting Ankara has been to keep up with the transnational patterns taking place in Western Europe. Turkey signed an association agreement with the then EC in 1963. Its objective in establishing links with the EC was to enhance its European identity. Full EU integration would legitimize and thus stabilize the various co-operative transactions with Western Europe. But two basic problems have stalled Turkey's 1987 formal application to join the EU: the issue of Cyprus and Kurdish human rights. Turkey is also identified as an essentially agricultural and underdeveloped country, that would drain the reserves of the Common Agricultural reserve, the European Investment Bank, and the Cohesion Fund. Realizing that closer relations with Western Europe are being hampered by the Cypriot affair, Turkey has entered a process of dialogue with Greece under the supervision of the UN and EU to find a solution to this Island's divide. Ankara has also signed a customs union with the EU and has actively been promoting transnational economic, financial and cultural links with Western Europe.

Another area which supports Turkey's institutional ties with Europe is its strategic location. Turkey's role in the 1990-91 Gulf War re-established its position as an essential component in the western security alliance at precisely the time when the end of the Cold War forebode that Turkey could become a less significant player in the European region (Sezer, 1992: 122-6). In line with the other members of the Southern European subgrouping, Turkey shares the post-Cold War conviction that it risks being marginalized by the core sector of the Western European international region, which is increasingly showing signs of establishing stronger ties with Central and Eastern European countries. If such a pattern of relations was sought at the expense of Southern European relations in general, and Turkish relations in particular, Turkey might be forced to reconsider more closely the level and nature of its relations with the Middle East. Ankara's dependence on Arab oil and growing support for Turkish fundamentalism are already pressuring the country to continuously reassess its position vis-à-vis this region. A shift to a more Middle Eastern dominant foreign policy would naturally affect Turkey's standing in Europe and could theoretically eliminate it from this international region altogether (see Buzan, 1991a: 215-21).

Like its Southern European counterparts, Turkey is also concerned with similar post-Cold War security challenges. These include ideological confrontations between Islamic fundamentalists and secularists in North

Africa, nationalistic and ethnic confrontations in the Balkans, and economic frustrations in both subzones which risk triggering international terrorism and massive population movements throughout Southern Europe. The key difference between Turkey and its Southern European neighbours is that while none of the others have direct territorial contact with the areas that breed potential threats to peace, Turkey does. Turkey is also keen to maintain secure access to the Sea Lines of Communications in the Mediterranean. Yet, Turkey's exclusion from the EU and its conflict with Greece risks driving a wedge between it and the rest of the countries in the region and keeps it in the peripheral sector of the Southern European subgrouping (Sezer, 1992: 132-33).

Portugal

Another candidate in the peripheral sector of the Southern European subgrouping is Portugal. It is included largely because of its decision to balance Western European foreign policy priorities with its Atlantic, Latin American and African external dimensions. Although Portugal is a member of both NATO and the EU, it has maintained close links with ex-colonies and Portuguese speaking countries such as Angola and Brazil.

Portugal maintains strong security relations with the United States and both countries have remained committed to their 1951 Defence Agreement. Lisbon's main benefit from this close relationship is that the US has regularly re-equipped Portuguese armed forces and provided a constant supply of military and economic aid under the Lajes airbase agreement (Santos, 1992: 89-91). Despite accession to the EU in 1986, Portugal continues to emphasize the importance of its trans-Atlantic linkage, in contrast to its neighbour, Spain, which in recent years has become an ardent advocate of establishing a common European foreign and security policy (see Table 4.6). Portugal's geographic position in the Atlantic also partly explains this difference in outlook. Thus, relations between the two Iberian countries are best described as cordial, with EU membership acting as a catalyst to an improvement in bilateral relations. One consequence of closer ties with Western Europe is that transnational ties with the rest of the continent have increased dramatically, as illustrated in the jump in trade with Spain for example, from 5% of total trade in 1985 to 15% in 1990 (see Agarwal et al., 1994, and Figure 4.2 and Tables 4.4 and 4.5).

Portugal's position in the Southern European subgrouping and the larger

Western European international region is perhaps best described as that of a balancer. Although it is committed to playing an active role in European Union affairs, it also regards itself as the champion of keeping the EU's external affairs agenda as transparent as possible. Thus, although it supports the integration of the WEU into the EU, it does not perceive this security arrangement as exclusive from either NATO or the OSCE. Lisbon also shares the common Southern European concern of being marginalized as a result of the end of the Cold War. It therefore argues that EU support for Eastern European countries should not overshadow the solidarity that has been shown to Africa, Latin America and the Middle East.

Issue	Spain	Portugal	Italy	Greece
Codecision for the European Parliament	In favour	Opposed	In favour	In favour
Federalism	In favour	Opposed	In favour	In favour
Principle of Subsidiarity	In favour	In favour	In favour	In favour
Majority Vote in CFSP	Ambivalent	Opposed	Ambivalent	Ambivalent
Inclusion of defence in CFSP	In favour	Opposed	Ambivalent	In favour
More cohesion money	In favour	In favour	In favour	In favour

Table 4.6 - Southern European Countries' Positions on Certain European Union Issues

Source: Holmes, John W. (ed.), *Maelstrom, The United States, Southern Europe, and the Challenges of the Mediterranean* (1995), World Peace Foundation, The Brookings Institution, p. 139

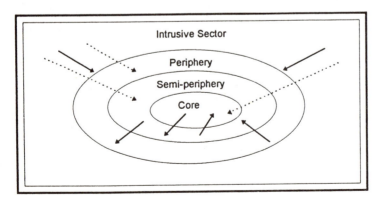

Figure 4.1 - Components Highlighting the Fluid Nature of International Subregional Groupings

Definition of Terminology

Core: Group of States which form a central focus of international politics within an international region.

Semi-periphery: Active participation within a given international region whose aspiration is membership of the core.

Periphery: States within an international region which are alienated from the core sector in some degree by social, political, economic or organizational factors, but which still play a role in the politics of the region.

Intrusive Sector: Politically significant participation of external powers in the international relations of an international region.

In the last decade Portugal has therefore expanded its European dimension but not at the expense of its trans-Atlantic and extra-continental ties. Its position in the Southern European subgrouping would have to be upgraded were it to commence dedicating more of its diplomatic resources towards European concerns. In the interim, Portugal already shows signs of moving in this direction by strengthening its relations with Maghreb countries, not only through participation in the Western Mediterranean Forum, but also through development of bilateral relations with Morocco and Algeria (ibid.: 92). It is also preparing to participate in UN peacekeeping missions for the first time.

Malta and Cyprus

The islands of Malta and Cyprus have been relegated to the peripheral sector of this subgrouping mainly because they have not yet been successful in their quest to become full EU members. As a result, their influence in regional affairs has been somewhat limited. Both islands are located in sensitive geo-strategic positions and could potentially play proactive roles in their respective spheres of influence. As already evident from their negotiations with the EU, both Mediterranean islands are keen to shed their neutral and non-aligned movement image and to begin playing a more proactive role in international security. Both countries have also become keen advocates of establishing trans-Mediterranean security initiatives, supporting the concept of a CSCM. The Maltese have also proposed the setting up of a Council of the Mediterranean, along the lines of a Council of Europe (Calleya, 1994a: 138-47), and have initiated a drive to create a Euro-Mediterranean stability pact, a concept endorsed by the French at the Barcelona Conference in November 1995 (Calleya, 1996: 9).

From an economic perspective, both countries are already deeply entrenched in the European sphere of influence, conducting more than two-thirds of their trade with the Western European international region. Admittance to the EU - which could take place as early as 1998, after the EU 1996 intergovernmental conference - would therefore require a reassessment of Malta's and Cyprus's position in the Southern European subgrouping. One likely outcome is that either one or both of the islands find themselves in the semi-peripheral sector of this subregion, commensurate to the increase of their authority in the area.

The liberation of Central and Eastern Europe, the demise of the Soviet Union, and the measures introduced to further the European Union experiment have forced the countries of Southern Europe to reassess their specific role as a Mediterranean subzone and to ponder what impact post-Cold War changes will have on their identity.

Interaction between the countries of Southern Europe is largely co-operative dominant. Exchanges are both intergovernmental and transnational in nature. However it is essential to note that patterns of relations are far more co-ordinated at a Western European level than at a Southern European one. In fact, EU member states in this subgrouping appear to be gradually moving towards a comprehensive type of regional arrangement with their northern neighbours. The EU can also be credited for fostering strong intergovernmental and transnational economic, political and social ties between these countries. The security dimension is however less durable in this subgrouping. Although members of the same political or military organizations (NATO, the WEU, or the OSCE), Southern European countries do not adhere to a common defence identity. France, Italy and Spain have recently supported the WEU more fervently, but Turkey (not a member of the WEU) and Portugal continue to favour NATO (see Table 4.6). On the other hand, Malta and Cyprus have favoured trans-Mediterranean security initiatives. In any case, none of these Western security institutions have been able to extend their security guarantees to the Middle Eastern subgroupings bordering the Mediterranean. As a result, Southern European patterns of interaction are perceived by states in the Levant and Maghreb as largely divisive mechanisms. Overtures to overcome this perception have to date failed, with the most successful attempt, the Western Mediterranean Forum, currently in a stalemate situation.

Southern European interaction with countries outside the Mediterranean theatre reflects the lack of foreign policy convergence among this subregional grouping. For example, after bilateral relations with other EU member states, countries such as France, Italy, Portugal and Spain dedicate most of their diplomatic resources on four priority locations: European Union affairs, bilateral relations with the Maghreb, bilateral contacts with the Levant, ties with Latin America and relations with former colonies throughout Africa. Mediterranean issues are therefore far down the list of foreign policy concerns, despite rhetoric claiming the contrary. Interaction between these four Southern European countries and their external contacts is primarily of an economic and political nature. Military links are relatively limited. Social intercourse is much more restricted, due to legislation preventing the mass migration of Latin American and African citizens to the more affluent European continent. However, although Southern European external policies do not coincide, they do somewhat overlap (Vasconcelos, 1992: 23).

On a more positive note, Southern European countries do share a set of interests that have gained currency since the end of the Cold War. First, is the belief that their European subgrouping needs more binding social and economic cohesion. Southern European EU member states have consistently put forward political programmes which focus on issues in and around the Mediterranean area at European Union Council meetings. An assessment of the 1995 French and Spanish EU presidencies and a glance at the 1996 Italian EU presidential programme reflects this trend. A second unifying factor is the conviction that measures need to be introduced to bridge the large gap between the two international regions which border the Mediterranean, and specifically Southern Europe, the Maghreb and the Levant. Measures to quell the revival of nationalism and emergence of racism in the area are also urgently required (ibid.: 15). However, an overall assessment of Southern European foreign policy priorities reveals that Mediterranean affairs are far down their external agenda, despite rhetoric claiming the contrary.

4.3.2 Degree and Types of Interaction in the Levant

As described at the start of this chapter, the Middle East international region is divided into three basic geopolitical subgroupings: the Levant, the Maghreb and the Persian Gulf. Encompassing the eastern and southern sectors of the Mediterranean, it is the first two subregions that concern us in this review of patterns of interactions in the Mediterranean area.

The Levant label is used to represent those countries directly associated with the Middle East peace process: Israel, Lebanon, Syria, Jordan, Egypt and the Palestinians. Iran, Iraq and Saudi Arabia are excluded from this analysis as they are considered to be outside this catchment area. Turkey is also left out as it has been classified as primarily falling in the Western

European international region. In this section, the area of the Middle East coincides with both geographical and geopolitical interpretations. In this analysis, primary concern is given to those countries that either have direct access to the Mediterranean or else dedicate their diplomatic resources to Mediterranean affairs. These are Israel, Egypt, Syria, Jordan and Lebanon. More than any other area of a comparable size, the Levant has been characterized by intense interaction as well as extremely acute regional conflicts. There is a high level of interconnectedness between the Arab actors in this area at both a societal and a state level, a distinct characteristic among international regions which comprise developing countries. Transnational social patterns of interaction include the movement of intellectuals, students, and especially workers who take up residence in other Arab countries for varying periods (Noble, 1991: 55-7).

Applying the regional model outlined earlier assists in conceptualizing which countries in the area yield the most power in this Middle East subregion. The core states in this sector are Israel and Egypt. For most of the post-war period, they have dominated the regional system of states in the Levant. They are also superior to other nations materially, militarily, and in motivation (see Tables 4.4 and 4.5).

Israel

Israel's predominant position in the Levant is due to its specific character as an extraterritorial state (Korany and Dessouki, 1991: 41). In other words, as the only Jewish state, Israel is identified as the home of Diaspora Jews, including the six million Jews in North America who have provided Israel's basic political, economic, and financial support (Brecher, 1969: 137). Without this extensive international network of alliances, Israel's existence, let alone its presence in the core sector of the Levant, would have been much more difficult to defend.

Throughout the Cold War, Israel's pattern of relations in the Levant has been predominantly conflictual. Intergovernmental relations did not officially take place until Egypt broke ranks with its fellow Arabs and recognized Israel in 1979. Transnational contacts were also limited to military patterns of interaction, mainly during the 1948, 1956, 1967 and 1973 Arab-Israeli wars. Transnational links in other sectors were largely non-existent.

The September 1993 peace agreement with the PLO and the October 1994 peace treaty with Jordan have finally paved the way for more co-operative transnational activities to take place in the Levant, particularly between Jordan and Israel. Joint business ventures between Tel Aviv and Amman have already been drawn up, with agreements reached in the telecommunications and financial sectors. Foreign direct investment in this subregion is also gradually increasing, with the tourist sector the largest growing industry in the short-term. The EU is the international organization which has offered the most investment capital to help develop and standardize neighbouring infrastructures. One project which is already underway is that which seeks to build a trans-Levant highway, joining Cairo and Riyadh in the South with Amman and Tel Aviv in the North. Such links are necessary if commercial and social interaction between these countries is to expand. In short, these limited co-operative intergovernmental and transnational interactions between the core states of the Levant and parts of its periphery bolster the relative position of Israel in this subgrouping. It is, however, too early to assume that Israel will, as a result, be able to play a more proactive role in non-Levant affairs. To date, efforts to establish non-regional ties have mainly been directed towards forging stronger ties with Asia and not with countries in the Mediterranean. The immediate concerns of Israel continue to be Levant based. Its diplomatic activities continue to concentrate on trying to build a strong co-operative network of relations with Egypt and Jordan, and strengthening its embryonic ties with the PLO. It is also seeking to nurture the rudimentary links it has established with other Arab countries in the Middle East, particularly Morocco and Tunisia. Given the fact that the economy of Israel is bigger than that of Egypt, Syria, Jordan and Lebanon combined, any steps towards establishing a free trade area in this subregion will reinforce the position of Israel as the centre of gravity in the Middle East international region (International Herald Tribune, 16-2-95: 8).

Egypt

The most assertive of the Arab states in the 1950s and 1960s was also the most stable: Egypt. During the leadership of President Gamal Abdal Nasser, Cairo emerged as the capital of the Arab world. Nasser adopted a foreign policy orientation which identified Egypt as an Arab actor functioning within three basic rank-ordered "circles": Arab, African, and Islamic. Egypt's concentration of power stemmed in part from its vast superiority in material capabilities (see Tables 4.4 and 4.5). In the 1960s, Egypt possessed almost half of the military capabilities of all the Arab actors in the Levant, with more than a 2:1 advantage over its closest competitor, Iraq. Economic power was more evenly spread, with Egypt's large population the second lowest per capita income in the area (Noble, 1991: 62). The absence of a Mediterranean circle is understandable because it would have meant co-operative interaction with European Mediterranean countries and engagement in Cold War battles (Selim, 1994: 6-7).

During the 1970s and 1980s, Arab patterns of relations began to resemble more of an unbalanced multipolar system. Economic power began to play a far more important role than military power, mainly due to the importance attached to Arab oil. The large pool of surplus capital in the region led to a more co-operative pattern of relations among Arab states, with wealthy oil-producing states much more prepared to assist their weaker Arab neighbours. A lack of leaders like Nasser, also resulted in the declining role of transnational political influence across the region. It was during this period that Syria experienced a significant increase in military capabilities and for a time occupied a position in the core sector. By the mid-1980s it had emerged as the largest Arab military power, slightly eclipsing Iraq and Egypt.

Anwar Sadat's recognition of Israel relegated Egypt out of the Arab world mainstream. One important consequence from its peace with Israel was improved relations with the Western world, particularly the United States, which started to furnish Cairo with regular military and civil aid. After a decade of enforced isolation, Cairo increased its activities in the Arab sphere in the latter half of the 1980s. By 1985, a recurring theme in Egypt's foreign policy is that which emphasizes the inter-connections between Europe, the Mediterranean, and the Middle East (Ajami, 1987: 108-38). Egypt's stature as a principal actor in the Arab world was officially restored at the Arab League Casablanca Summit in May 1989.

President Mubarak set the tone of Egypt's foreign policy in the nineties in an address to the European Parliament in November 1991 when he suggested the creation of "a Mediterranean forum which will increasingly expand to include all European and Middle Eastern nations" (Selim, 1994: 6-7). In addition to its crucial role in the Arab-Israeli peace talks, Egypt has therefore sought to enhance its regional positioning by again trying to breathe life into the notion of a trans-Mediterranean security forum. Egypt introduced this shift in its foreign policy because it realized that the end of the Cold War created new opportunities and, simultaneously, new risks for Egypt. Fearing marginalization by Western Europe now that Central and Eastern Europe had emerged from behind the Iron Curtain and Southern European countries had excluded it from the 'five plus five' process, Cairo saw the Mediterranean policy as a mechanism through which Western European interests in the eastern sector of the basin could be strengthened. Moreover, the Mediterranean dimension also offered Egypt the opportunity of diversifying its external economic and political relations and not run the risk of becoming a U.S. pawn in the area. Egypt outlined the extent of interaction it supports in the Mediterranean at the Ministerial Meeting of the Nucleus Group of the Mediterranean Forum held in Alexandria in July 1994 (Calleya, 1994d: 33; see also Selim, 1994: 15-17). Discussion focused on issues of energy, science and technology for development, and environmental protection. The Mediterranean Forum which the Egyptians are proposing can largely be labelled a technical institution which concentrates on channelling European capital to the southern shores of the Mediterranean. In contrast to France, Italy and Tunisia, Egypt is against the granting of a mandate in security matters (Selim, 1994: 16).

Thus both Israel and Egypt have consistently manifested their "core" credentials by demonstrating an ability to influence balance of power dynamics within their own region. The efforts of Israel to keep the peace process going and the endeavours of Egypt to adopt a more vocal Mediterranean foreign policy orientation are both attempts to preserve their respective positions of strength in the Levant.

Syria and Jordan

The semi-peripheral states in the Levant consist of Syria and Jordan. Their level of power permits them to play limited and selected roles in and around their own region. Unlike the core states, the semi-peripheral states tend to be more unpredictable when conducting their foreign policy, more ideological and less industrialized. Syria's links with the Soviet Union and its spoiler attitude in the recent Middle East peace talks demonstrates this type of foreign policy behaviour. During the 1970s, an improvement in Syrian domestic political conditions and an increase in military reserves propelled the country to the forefront of relations in the Levant. Yet President Assad's differences with Egypt and Iraq and his lukewarm relations with Saudi Arabia soon eliminated Syria from the Levant core system. By the late 1980s, Syria faced the double blow of having to live with a victorious Iraq on its eastern flank and the withdrawal of Soviet assistance (Hinnebusch, 1991: 374-409).

The distinctive characteristic of this sectoral division is that the countries classified possess enough power to entice external actors to assist them in their independent regional ambitions. Throughout the Cold War, Syria was one of the main client states of the Soviet Union in the Levant. In the nineties, it is evident that a lasting Arab-Israeli peace cannot be brokered without the inclusion of Syria in a comprehensive peace agreement. In consequence, the United States has recently dedicated a large proportion of its diplomatic efforts in the Levant in trying to attract Syria towards the negotiating table. Seen from such a perspective, post-Cold War regional developments have had an ambiguous influence on the position of Syria in this subregion. This ambiguity stems from the fact that Syria is a key player in the Middle East peace process, yet faces marginalization because of its rejectionist stance.

The limited power projection of Jordan is the main reason for its inclusion in the semi-periphery of the Levant. Its decision to sign a peace agreement with Israel in 1994 exhibits the limitations of its regional influence. Once the Palestinian Liberation Organisation (PLO) followed the route taken by Egypt a decade earlier, Amman had little choice but to follow suit, or else risk being isolated from the main pattern of relations functioning in the area. Such an outcome would have relegated Jordan from the semi-periphery to the periphery of this subgrouping. Despite joining the peace process bandwagon, Jordanian relations with its Arab neighbours remain erratic. The opposition of King Hussein to the war against Iraq drove a wedge between Amman and the rest of the Arab states that supported the UN coalition of forces. As a result, Jordanian relations with Egypt remain distant, while links with Saudi Arabia, which it accuses of meddling in Yemeni internal affairs are even more limited (Financial Times, 10-6-94: 19).

After a series of events which saw Amman lose diplomatic ground in the last few years, a few factors indicate that the relative position of Jordan in the semiperiphery of the Levant could begin to improve in the short-term. One immediate benefit from the signing of the peace agreement with Israel in October 1994 is the rapid increase in cross-border activity between these two countries. The co-operative intergovernmental action between Jordan and Israel has triggered a plethora of transnational forces which include joint infrastructural programmes, co-operation in the financial sector and a rise in foreign direct investment. Given its strategic geographic location, Jordan is already devising plans to be able to act as a transfer point for commodities arriving from the West and destined for the Gulf and Saudi peninsula. King Hussein has also introduced the first steps towards developing a democracy with the first multi-party elections held in 1993. The political stability of Jordan coupled with an increase in its domestic and international economic activities such as hosting the Middle East/North Africa economic summit in

October 1995 assures that Amman will continue to play an active part in the Levant's semi-periphery. If regional peace flourishes and Jordan is successful in executing its economic and financial programmes, the country could begin to play a more significant role in Mediterranean affairs. However, in the short-term, Jordan will be caught up in investing all of its diplomatic resources in consolidating the Middle East peace process.

Lebanon and the Palestinian entity

The core and semi-peripheral states are the main players in a regional system. In contrast, peripheral states are much more passive, and are so weak they can only react to the international relations of their region. In this case, the periphery includes Lebanon and the Palestinian entity. These actors have very little or no influence in regional affairs. They are often within a regional powers' sphere of influence or are completely side-lined by the more influential actors. In areas where geographic position, material weakness and internal political chaos have combined, a few peripheral states have emerged as battlegrounds, e.g. Lebanon. The balkanization of Lebanon has eased in the nineties but the country remains to all intents and purposes a puppet of Syria. Palestinian representatives have also made a few significant steps forward following the agreement signed between Israel and the PLO in 1993. They however remain in the periphery of Levant interaction, largely dependent on other actors to help them voice their concerns.

During the post-Cold War period, interaction between the countries of this subgrouping is no longer conflictual dominant. The aftermath of the 1990-91 Gulf War has seen a shift in regional alignments and ushered in a new core coalition of leading powers in the area which includes Egypt, Saudi Arabia, and potentially Syria. The prospect that Arab co-operation can prove durable, remains doubtful. Yet, the 1993 Israeli-Palestinian breakthrough has had two main effects in this respect. First, it has strengthened the relative position of all the states who are playing an active role in the peace process. Greater domestic stability permits these states to exercise greater flexibility in their foreign policy actions. Secondly, the links established between Israel, Egypt, Jordan and the Palestinians are serving as a foundation upon which working relationships, perhaps of even a transnational nature, can be built across previously rigid ideological divisions. In the interim, intergovernmental patterns of relations continue to guide contacts in the Levant. The key difference is that connections are now co-operative dominant. On the other hand, the resurgence of Islamic fundamentalism, particularly if it were to make substantial inroads in Egypt, could easily push this delicate network system in the opposite direction, with an increase in the enmity variable.

Paradoxically, while prospects for economic links are improving, political ties among the Arab actors in this subgrouping are weakening. An important contributing factor is the decline of pan-Arabism (Ajami, 1978: 357-69: Ajami, 1981). This decline in the intensity of identification with the larger Arab community has also led to a reduction in commitment regarding Arab core issues. As a result, pan-Arab nationalist movements have weakened, fragmented or disappeared. Nevertheless, Arab societies and political systems remain interconnected and permeable to some degree (Noble, 1991: 60). In the last few years, there has been a clear increase in transnational responsiveness and cross-frontier ties based on Islam. The establishment of an Islamic state in somewhere like Algeria could act as a catalyst, and increase the density of cross-border interactions throughout the Middle Eastern international region.

While a diffusion of power has contributed to a more fragmented Levant subgrouping, historical and religious legacies continue to dictate the functioning of an intense pattern of relations. This is evident when one assesses the foreign policy priorities of the actors in this area: bilateral regional relations remain top of foreign policy agenda. These are followed by an urgency to establish strong relations with the last superpower, the United States. Bilateral contacts with Western European states and links with the European Union have recently gained in prominence, with Mediterranean issues figuring further down the foreign policy priority chain.

A tentative way of describing the current network of relations in the Levant is to say that they are much less revolutionary than before. The more competitive multipolar structure at a subgrouping level, an international region level and a global level, prevents any one actor from exercising any hegemonic ambitions it may harbour. The moderation in relations within the Levant during the post-Cold War era has however not resulted in a diminishing of the intensity of relations. Israel's security concerns still stand in the way of implementing a free trade zone in this subregion. Several observers also believe that the absence of a solid agreement between Israel and the Palestinians or a comprehensive Levant peace embracing Syria and Lebanon, and continuing Arab fears of Israeli economic domination weakens any notion of Middle East integration (Financial Times, 9-2-95: 9). As a result, there is little scope to expect any of these actors to be in a position to contribute substantially to trans-Mediterranean security initiatives in the short-term, even if Egypt is already proving the exception.

4.3.3 Degree and Types of Interaction in the Maghreb

The term Maghreb describes those countries that are located along the western coastline of North Africa. This consists of Algeria, Morocco, Libya and Tunisia. The concept of a Maghreb subgrouping adopted in this section is thus a geopolitical one (Spencer, 1993: 5-7). The core states in this sector are Morocco and Algeria. Events in one of these states nearly always has an impact on the other, despite their objective differences. In addition, their cultural composition of similar, yet distinct blends of Arabo-Berber, Sunni Muslim, and African and European traditions help create intense patterns of co-operation and conflict. These patterns of interaction are political dominant, not social or geohistorical like those in the Levant (Zartman, 1984: 151). In fact, whereas the network of relations in the Southern European grouping is co-operative dominant, and the links in the Levant have until very recently been conflict dominant, relations in the Maghreb have been much more mixed, with phases of conflict often being replaced by those of co-operation.

One factor that helps to explain the intensity of relations between the states in the Maghreb subgrouping is the complementarity that exists among these countries. Rather than making it easier for them to co-operate, their infrastructural similarities make them arch-competitors. For example, Algerian industrialisation is predicated on Algerian dominance of the Maghreb market, a policy that cannot work as the other states in the area attempt to develop their own industrial base (ibid.: 156).

Algeria and Morocco

Algeria and Morocco dominate the political, economic and military patterns of relations within this Middle Eastern subgrouping. Since achieving independence, these two countries have consistently attempted to tilt the balance of power, within this region, in their favour. In 1963 both clashed in a short-lived desert war over their inherited borders around the Western Sahara. In the past two decades Algeria, with the periodic support of Libya, has backed the Polisario Front in its struggle to assert the independence of a state - the Saharan Arab Democratic Republic (SADR). On the other side, Morocco has been attempting throughout to re-integrate this pre-colonial territory, that was dismembered by the Spanish and French, back into Morocco. Rabat has consistently been supported financially by Saudi Arabia and periodically by Mauritania. More recently, Mauritania has joined Tunisia in taking a neutral position over the issue (Spencer, 1993: 43).

Since 1991 the United Nations has managed to enforce a cease-fire in the area, but it has failed to enact a territorial referendum or otherwise propose a settlement that will bring long-term peace to the Western Sahara. Morocco has been able to sustain a favourable military supremacy for itself which has gradually eroded Polisario's military, and more importantly political objectives. It has also consolidated its position through population transfers to the Western Sahara and by increasing administrative control over Saharan population centres. Despite periodic improvements in relations between Morocco and Algeria through forums such as the Arab Maghreb Union, the latter continues to provide financial assistance and territorial sanctuary for Polisario military forces. Thus the Western Sahara conflict has the potential to continue as a low-intensity military confrontation throughout the 1990s (Bauer, 1994: 111-29, and Spencer, 1993: 44-5). This conflict seems set to follow the usual African pattern of brief hostilities until the exhaustion of resources, followed by a cease-fire and then a lengthy period during which both progress and regress in the outcome of the conflict is registered (Zartman, 1984: 153).

Trends in the last decade indicate that Algeria's position in the core sector of this Middle Eastern subgrouping has been constantly slipping. The collapse of oil prices in 1986 forced Algeria to reassess its policy of the 1970s and 1980s of regional hegemony. Algeria has also been profoundly weakened by its internal crisis over the future form of government it will have and over its full-scale confrontation with Islamic movements operating within its territory. These factors exacerbate further the fact that Algeria's armed forces are much smaller than those of Morocco, placing further constraints on Algiers' regional strategic ambitions. As part of the Arab Maghreb Union, which it formed in February 1989 with Morocco, Tunisia, Libya, and Mauritania, Algeria agreed to a mutual defence pact, which places a further restraint on any military pretensions it might have in the region (Joffe, 1994: 25-7). Algeria's failure to clamp down on Islamic militants, as vividly demonstrated by the Christmas 1994 highjacking of an Air France 737 at Boumediene airport, and its insufficient managing of its governmental crisis, indicate that its role in the core sector has become that of a maverick rather than the regional stabiliser that it was in the past. In contrast, Morocco's position seems to have been relatively strengthened. Its high profile hosting of the Middle Eastern economic summit in October 1994 testifies to its leadership aspirations not only in the Maghreb, but throughout the Middle East international region.

Libya and Tunisia

The semi-peripheral sector consists of Libya and Tunisia. Libya's level of power permits it to play a limited and selective role in and around this geopolitical area. Libya fits into the characteristics of semi-peripheral states described previously. Tripoli has constantly displayed its ability to act unpredictably when conducting its activist and revisionist foreign policy, and it has demonstrated a capability to be more ideological than its neighbours on numerous occasions (ibid.: 27-8). Libya is also less industrialized than either Algeria or Morocco. However, the distinctive feature that illustrates Libya's semi-peripheral status which occasionally progressed towards the core position, is its ability to entice more powerful states to become involved in this subgroupings' network of interactions. The Soviet Union's Cold War track record gives clear credence to this fact. Moscow supplied the Gaddafi regime with both economic and military assistance throughout the 1970s and 1980s in exchange for access to the warm water ports scattered along the Gulf of Sidra. The 1986 U.S. bombing of Tripoli is another example of this ability to attract great power attention to the Maghreb.

Of all the countries in the Maghreb, only Libya's hostility to the West is endemic. It possesses a variety of short-range ballistic missiles which are capable of reaching Malta, Sicily and Southern Italy (Farley, 1994: 34). It also has a submarine and air capacity which might be in a position to harass Western shipping lines in the Western Mediterranean waterways. Given its size, it is perhaps the most militarized country on the North African littoral with 2,300 tanks, as against Morocco's 250 and Algeria's 900 (*The Military Balance, 1994-95*, IISS).

The post-Cold War period has again witnessed the ability of the Libyans to maintain superpower and great power interest in the area, although now for different reasons. Although the Lockerbie affair has generally had a negative impact on Libya's economic development, it has not affected the country's main hydrocarbon export industry. From a political perspective, the whole affair can be viewed as actually enhancing the position of the Gaddafi regime internationally. The United States and the West in general seem resolved to accept that they are better off dealing with the devil they know rather than taking the risk of negotiating with less known political authorities that could emerge if a power vacuum were to develop in Tripoli (Calleya, 1994d: 33).

In this sense, Libya is typical of a number of post-Cold War states whose internal dynamics present potential, but not yet realized, external security risks. At best, Western governments keen to promote stability in the western Mediterranean, have one foreign policy option that will help avoid exacerbating the problem. Beyond economic support, external actors must find a way of bringing Tripoli back into the international society of states, if conflictual patterns of interaction are to be reduced in this sector of the Mediterranean. Failure to nurture co-operative links with Libya could result in repercussions that would extend beyond the subregion itself. These risks include an accelerated flow of refugees seeking political asylum, (especially to Italy), an exacerbation of racial and religious tensions both within North Africa and Western Europe and between Western Europe and North Africa, and the likely exportation of terrorism similar to that of the 1980s.

By contrast, Tunisia is primarily in the Maghreb semi-peripheral sector because of its sheer size in demographic, economic and military terms. With a population one third of that of either Algeria or Morocco and a falling birthrate faster than either, Tunisia has registered phenomenal growth figures in the past two decades under the stewardships of Bourguiba and, after 1987, Ben Ali (Zartman, 1984: 161). This is partly explained by the fact that the Tunisian economy is the most open to foreign investment in the area, a position it has held since the early seventies (Spencer, 1993: 17). This puts Tunisia in a better position than its neighbours in relative terms to contend with its international debt which spiralled mid-way through the 1970s (ibid.: 18-19 and 38-9).   In military terms, Tunisia is no match for either Algeria or Morocco. The former has an army four times larger and an airforce six times larger than that of Tunisia, while the latter has an advantage of 6:1 and 3:1 respectively (see Table 4.5 for Maghreb military figures, *The Military Balance, 1994-95*, IISS).

Its foreign policy track record is much more consistent than that of its neighbours, opting to remain neutral in most cases of regional conflict. This policy, ironically, promoted Tunisia to the front line of diplomatic activity in the Middle East. In 1978 the Arab League's headquarters were moved to Tunis after Egypt was excluded from the League's proceedings having signed the Camp David Accords. In 1982, the PLO headquarters were shifted to the outskirts of Tunis following the Israeli invasion of Lebanon. However, Tunisia's reluctance to support the UN coalition in the Gulf crisis of 1990-91 seemed to start a process which has seen it lose some of its influence in the Middle East international region. Income from tourists and the cessation of exports to both Iraq and Kuwait were a considerable loss for the Tunisian economy. Even more unpleasant is the suspension of generous financial aid from the Gulf monarchies and the dramatic reduction in American aid (Biad, 1993: 28). Recent improvement in relations among countries in the Levant have also somewhat reduced Tunisia's fortunes, with the Arab League headquarters back in Cairo and the PLO now located in Jericho.

Thus, while the Middle East continues to exercise its historical and cultural hold over Tunisia, its distance from the main areas of conflict within this international region have made it marginal to their resolution. By contrast, Tunisian links with the Western European international region are steadily increasing. In addition to economic partnership negotiations with the EU, Tunisia has signed a series of bilateral agreements with Italy which will increase its energy output across the Mediterranean. In March 1991, Italy and Tunisia agreed to double the capacity of the 1983 gas pipeline which stretches across the Straits of Gibraltar to 40bn cubic metres by the year 2000. Tunisia is also to start exporting gas-generated electric power to Italy by the mid-1990s (Spencer, 1993: 52).

An assessment of the patterns of relations between countries in the Maghreb subgrouping therefore reflects clearly their co-operative and conflictual nature. Interaction is predominantly intergovernmental, with transnational intercourse at an embryonic stage. For example, intra-Maghreb trade has barely increased beyond 1.5% of their total exchanges since independence (ibid.: 47). An increase in cross-border activity has been registered in the area of Islam, with fundamentalist groupings such as the Muslim Brotherhood supplying basic medical and educational services at a grass-roots level to help make up for governmental inadequacies in these sectors. An analogy which accurately depicts interaction in the Maghreb is that of the checkerboard pattern of relations. According to this basic configuration of international relations, the states of the region have consistently acted as if "my neighbour is my enemy and my neighbour's neighbour is my friend". This is the pattern of relations that was encouraged by the French during the nationalist period and it is the type of contact which has characterized relationships in the region thereafter. Even institutionalized attempts at regional unity, such as the Organization of African Unity, the Arab League and the Arab Maghreb Union (AMU), have been dominated and broken by this same pattern (Zartman, 1984: 159).

There is very little to suggest that this trend will change in the post-Cold War. Foreign policy considerations remain as divergent as ever, as highlighted in AMU proceedings held during the Gulf War. Morocco approved of a resolution "condemning the aggression against Kuwait, and Iraqi threats against the Gulf States", Libya rejected it, Algeria abstained, while Tunisia boycotted the summit (Biad, 1993: 25).

Priorities in external affairs also remain varied. While bilateral relations in the Maghreb continue to receive special attention, the Algerian crisis and Libyan isolation have stalled any effective Arab Maghreb Union resolutions from taking place. At an AMU summit at Ras-Ahnouf in Libya in 1992, a Nine Point plan of regional co-operation was initialled, stipulating the creation of a free exchange zone by the end of 1992, the creation of a customs union by the end of 1995, and the creation of a Common Maghreb Market by the end of 2000 (ibid.: 31 and 34). In reality, little or no progress has been registered in removing trade barriers within the Maghreb. As a result, both Morocco and Tunisia have concentrated their diplomatic resources on establishing stronger economic and political ties with the Western European international region. Both have signed partnership agreements with the EU which will permit them to participate openly in European economic and financial circles. Algeria and Libya remain on the sidelines of these negotiations because of their unstable domestic situations. Links with the rest of the world figure further down the foreign policy chain, including pan-Mediterranean issues (see Table 4.1).

Historically, North Africa has only been a political and economic unit on very few occasions. There is no concept of a "Maghreb nation" as there is of an "Arab nation" despite efforts to institutionalize such a concept. In fact, no single Arabic word exists for the term North Africa, since "Maghreb" refers properly to Morocco. The logic of reality thus points towards future relations in the Maghreb which are of accentuated conflict and interrupted co-operation, oscillating between these two poles, but never attaining either because of the multipolar division of power in the area. Recent intergovernmental political and economic links between some of the countries in this area (Morocco and Tunisia) with Europe could be interpreted as signalling the start of a period of cross-fertilization or homogenization between Western Europe and this Middle Eastern subgrouping. However, transnational statistics, apart from those concerning the energy sector, reveal that this is unlikely to be the case. It should also be noted that interaction and interdependence do not necessarily imdicate the emergence of a co-operative international region. They can also create the ingredients of conflict. It is thus too early to speak of comprehensive bridges being built across the Euro-Maghreb divide, let alone of a process which might lead to the development of an entity resembling a co-operative or conflict based Mediterranean international region, that incorporates Southern Europe, the Levant and the Maghreb.

4.4 Southern Europe, the Levant and the Maghreb: Connections and Disconnections

Having examined the patterns of interactions within Southern Europe, the Levant and the Maghreb, how objectively correct is it to claim that a Mediterranean international region of some type is emerging? Are the patterns of interactions within and especially between these three subgroupings strong enough to suggest the formation of a Mediterranean international region? Do recent exchanges between the European and Middle Eastern subgroupings bordering the Mediterranean suggest the emergence of relations which are transnational, intergovernmental or comprehensive dominant? Or do relations across the basin constitute essentially distinct formations, each defined much more by what goes on within each subgrouping, than by interaction between the three subregions bordering the Mediterranean? A brief synopsis of the historical account provided in chapter three will assist in answering these questions.

Traditionally there is plenty that links these three Mediterranean areas. For almost a thousand years starting from Alexander the Great's empire until the end of the Pax Romana, the Mediterranean and its surrounding hinterlands formed part of a single political system dominated by a Hellenistic culture. Throughout the seventh and eighth centuries, the Mediterranean served as the backdrop for the bitter military struggles between Christian Europe and the Islamic world. The start of the nineteenth century saw the Ottoman empire decline to the point that they could no longer resist European expansion into the Middle East. This interest increased as oil became more of an important commodity by the turn of the century. The defeat of the Ottoman empire in the First World War resulted in the dismemberment of its Arab components into League of Nations' mandates and protectorates. The aspiration of Arab or Islamic unity was thus destroyed as the foundations for a number of separate states were laid.

Despite the long history of interlinkage, the gradual European withdrawal from the eastern and southern sectors of the Mediterranean opened up a window of opportunity for indigenous patterns of regional interaction to increase. Having exhausted itself in two world wars, the European space yielded to superpower overlay. In the interim, indigenous forces within the

Levant and the Maghreb sought to eliminate any imperial remnants from the vicinity. The Suez fiasco in 1956 and the expulsion of France from Algeria in 1962 severely undermined Europe's position in the Mediterranean. By the start of the seventies the United States and the Soviet Union had replaced Europe as the principal intrusive actors in the basin. Although France maintained significant links with its former colonies in the Maghreb, the Europeans' strategic role in the area was now mainly as allies of the United States.

Throughout Southern Europe national interests continue to shape different approaches to the Mediterranean area. Within the general EU framework of co-operation and assistance, Southern European states feel they can act somewhat autonomously. This for example, helps explain the occasional competitive nature between France and Spain in the area (Ortega, 1995: 43). At an EU level, Southern European states have demonstrated that they are increasingly prepared to pool their diplomatic efforts to help lobby for more economic assistance to the Mediterranean area. The states within this subgrouping therefore seem to be following a flexible dual-track policy towards the Mediterranean: an independent bilateral approach at a national level and a co-ordinated multilateral approach at an EU level.

The founding of the Arab League in 1944-45 is one of the first clear signs of indigenous patterns of regional interaction starting to gain ground in the Levant and Maghreb. The League expanded to over twenty members and covered a geographical area spanning from the Atlantic coast of North Africa to the Arabian Sea. A common Arab identity and Islamic religion were strong enough cohesive elements for a politico-cultural forum to develop. Pan-Arab nationalism and Islam created transnational linkages of a largely political nature throughout the Levant and the Maghreb. The sheer number of states and the diversity of issues that had to be addressed, guaranteed that this international region would have its fair share of alliances, rivalries and wars.

The question that dominated the Middle Eastern international region was the Arab-Israeli crisis concerning the Palestinians. Practically all of the Arabs (especially Egypt, Jordan and Syria) found themselves pitted against Israel with short wars occurring at frequent intervals (1948, 1956, 1967, 1973, 1978, 1982). The Middle East peace process breakthrough of 1993 augurs for a period in which indigenous regional dynamics will be able to evolve in a more co-operative manner. Interaction will remain intense as countries in the Levant compete for their share of foreign direct investment. Transnational and intergovernmental political and military linkages may eventually be supplemented by social and economic exchanges. If this is the case, this subregion could develop comprehensive patterns of relations in the long-term.

Further East, intense rivalry between the key actors in the Gulf, namely Iran, Iraq and Saudi Arabia, is sure to continue and thus risk upsetting co-operative patterns of relations throughout the Middle East. The withdrawal of Britain from the Gulf in the late 1960s allowed local regional dynamics to again develop unconstrained. The most fierce conflicts in the area to date are the eight year Iran-Iraq war of the 1980s and the 1990-91 Gulf War to oust Iraq from Kuwait. The Persian Gulf remains unstable as both the Iraqi and Iranian regimes face domestic challenges that could alter the balance of power equation in the area. Any outbreak of hostilities in any of these countries could easily spill over into the adjacent Levant subregion and as a result upset recent efforts to build co-operative relations across the Middle East.

In the Maghreb, less intense rivalries have developed amongst Algeria, Morocco, Libya and Tunisia. The principal flashpoint in the area is the Spanish Sahara where forces loyal to Algeria and those supported by Morocco continue to influence the succession struggle. The creation of the Arab Maghreb Union in 1989 between Algeria, Morocco, Libya, Tunisia and Mauritania, seemed to usher in a period where very limited intergovernmental social, economic, political and military links could develop. It was also thought that the strengthening of this type of interaction would act as a catalyst to transnational patterns of interaction, which are largely absent in the Maghreb. The Libyan-Lockerbie incident, the Algerian civil crisis, and the August 1994 Algerian-Moroccan diplomatic scuffle have all but halted the process of establishing a more intensive co-operative intergovernmental Maghreb subregion.

In short, different forms of regionalism are taking place in Southern Europe, the Levant and the Maghreb. It becomes easier to understand the different forms of regional development in each of these subregions if one first clarifies the various levels of regionalism that can take place. Gibb provides an accurate table of the varying levels of regional integration, from the lowest form of co-operation to the highest (Gibb, 1994: 24, see Table 4.7).

When applied to the Mediterranean space, this classification of regional trading arrangements reveals the multitude of different schemes operating in a geographically proximate area. Southern Europe is by far the most integrated of the three subregions encompassing the Mediterranean. As members of the EU and signatories of the Single European Act, the countries of this grouping have already established a free trade area, a customs union, and a common market between them. In other words, countries in Southern Europe have eliminated completely quantitative trade restrictions and custom's tariffs against each other's goods and have agreed to adopt uniform import tariffs and common quota restrictions vis-à-vis countries outside this subregion. In addition, customs union provisions are extended to include the free movement of the factors of production (goods, people, capital and services) within and across the borders of participating states (ibid.: 25-6). Together with their counterparts in Western Europe, Southern European states, apart from Malta, Cyprus, and Turkey, are currently mapping out strategies to introduce higher forms of regional trading arrangements, namely economic and eventually political union.

	Removal of internal quotas and tarriffs	Common external customs tarriff	Free movements of land, labour, capital and services	Harmonization of economic policies and development of supranational institutions	Unification of and political and powerful supranational institutions
Sectoral Cooperation	●				
Free Trade Association	●				
Customs Union	●	●			
Common Market	●	●	●		
Economic Union	●	●	●	●	
Political Union	●	●	●	●	●

Table 4.7 - Varying Levels of Regional Integration

In contrast, no such patterns of commercial interaction exist in the Maghreb or the Levant. In the former, the lowest level of regional trading exists, sectoral co-operation. This type of interaction occurs when co-operation is limited to well-defined sectors of production, such as oil and gas. Cross-border energy projects between Morocco, Algeria and Tunisia illustrate sectoral co-operation in the Maghreb. In the Levant, sectoral trading arrangements are just beginning to take off, as intergovernmental relations have improved. Although limited to a very few sectors, such as tourism, regional infrastructural projects which include Jordan, Israel and Egypt in the short-term, and Lebanon and Syria in the longer-term, are already being drawn up and will facilitate negotiating sectoral trading agreements in other areas. Whether this low form of co-operation can eventually be upgraded to perhaps a free trade or common market level of integration is an open question and depends largely on how successful the countries in both the Maghreb and the Levant are in sustaining peaceful regional relations. In the interim, the Middle East - North Africa Economic Summit organized by the Geneva based World Economic Forum and the New York based Council on Foreign Relations in Casablanca in October 1994 and the follow up conference held in Amman in October 1995 have demonstrated the plethora of obstacles that have to be overcome before higher forms of regional integration can be attained within each subregion and the Middle East international region as a whole (New York Times, 1-11-94: 8; see also Financial Times, 13-2-95: 14). A relative cooling in Middle Eastern regional relations at the start of 1995 is reflected in the outright rejection by Gulf Arab countries to participate or fund a regional Middle East development bank (International Herald Tribune, 14-2-95: 7), scheduled to open in Cairo by 1998 (MEED; 1995:7).

Focusing attention on the nature of interactions between Southern Europe, the Levant and the Maghreb, it is evident that no single recurrent pattern of interaction is discernible. Except for energy and armament ties with Europe, plus some residual intergovernmental cultural and political links, the post-independence Levant and Maghreb have evolved in their own distinctive ways. The process of Cold War overlay in Europe and that of decolonization in the Levant and the Maghreb has therefore seen the southern and eastern Mediterranean hinterlands become more introverted. Intergovernmental economic and military interaction has continued unabated (oil and weapons purchases) but intergovernmental political and social ties have diminished and transnational linkages remain limited to the energy sector. Although the Western European international region in general, and its Southern European subregion in particular, have recently dedicated a slightly higher proportion of their economic and political resources towards establishing stronger relations with their Middle Eastern neighbours, the level of political and economic interaction between both international regions remains small. When the classification of regional trading arrangements (Table 4.7) is applied to the level of commercial interaction taking place between Western Europe and the Middle East, the sectoral co-operation level best describes existing patterns of interaction. The concept of a free trade area between Western Europe and a few countries in the Middle East, namely Morocco, Tunisia, Egypt and Israel, remains a fallacy. Despite political rhetoric referring to such trading arrangements, preferential trade agreements are the furthest Western Europe has been prepared to extend levels of co-operation. Moreover, increasing socioeconomic disparities between Europe and the Middle East reveals the lack of convergence that has been registered during the past two decades in spite of a series of "global Mediterranean policies" launched by Western Europe. To summarize, an analysis of the degree and types of interaction in the Mediterranean area reveals that the disconnections between Western Europe and the Middle East far out measure the connections. This shows that current references to centripetal tendencies in the Mediterranean area exist more in theory than in practice.

4.5 Conclusions

The Mediterranean Sea itself acts as a physical boundary between the Western European and Middle Eastern international regions. During the Cold War the Iron Curtain served this purpose. However, the demise of the Soviet Union has seen the notion of a European international region expand to new frontiers. Both Central and Eastern Europe now regard themselves as prospective candidates of such a grouping. When one moves further east towards the Baltic States and the former republics of the Soviet Union the density of interaction with Western Europe becomes less intense. As mentioned earlier, although classified as a peripheral actor in the Western European international region, Turkey also risks being caught between the two completely different international regions encompassing the Mediterranean. Seen from a more optimistic perspective, Turkey regards its membership of NATO and efforts to join the EU as strong enough credentials to qualify it as a European secular state. Islamic, but not Arab, Ankara aspires to act as an insulator between the distinct European and Middle Eastern international regions. Accession into the EU would certainly signify that Ankara has moved closer to the European orbit of relations in an economic and political sense, but its geographic position will continue to dictate that it remains linked in both transnational and intergovernmental terms to the Middle Eastern pattern of relations.

Examining links between these two international regions reveals that two patterns of interaction occur frequently at an international intergovernmental level in the field of "high politics", although it is not entirely clear what triggers these patterns to occur or why they sometimes appear to function simultaneously. These types of interaction are the symmetry pattern and the contrast pattern.

The symmetry pattern occurs when international regions interact in such a way as to produce similar processes within both areas. The symmetry pattern is observable when an "external threat" situation exists. This pattern of interaction underlines the idea that international regions tend to reinforce each other if

* there exists a mutual involvement between the societies of the two regions
* regional processes within one area are detrimental to the interests of the other
* there is no real possibility of participation in the other international region.

Once again the "threat effect" constitutes the main transmission belt. In less abstract terms, this hypothesis suggests that the success of one international region is harmful to the development of another similar system. This effect forces the latter to co-ordinate policies among national systems and to arrive at a consensus on how to maximize their comparative advantages (Kaiser, 1968: 100).

The symmetry pattern of interaction is apparent in Western European relations with the Levant and the Maghreb. Despite verbal rhetoric declaring that the area is of primary concern to the states of Western Europe and international organizations functioning in the area, such as the EU, and the high level of trade between the two international regions, Western Europe is reluctant to establish strong comprehensive ties with countries in these two Middle Eastern subregions.

The second type of interaction is the contrast pattern of interaction. International regions interact in such a way as to produce different, if not opposite processes within each region. Interacting regional subgroupings weaken the potential that may exist toward moving in the direction of any of the international region modalities if

* they hurt the interests of members of other international subgroupings
* they are reasonably strong and offer rewards for participating within the area
* participation in or access to the benefits of the alternative international region is an option that is reasonably open to individual members of the other regional gatherings.

For example, one can argue that the European Union has weakened the prospects of other regional movements in the area, such as the European Free Trade Area (EFTA) or the Arab Maghreb Union because of the opportunities which exist which allow members of these two gatherings to benefit from the comprehensive regional setting in Europe. Thus, by offering sufficient enough rewards to prospective candidates and remaining "open" enough to make participation through membership, association, or special arrangement a feasible alternative, the European Union has weakened other regional initiatives in its vicinity.

The existence of such an "open" international region which offers benefits to members of other international regions through special arrangements leaves the states with two alternatives to consider: they can either strengthen their own subgrouping in an effort to upgrade the patterns of relations within their particular international region by introducing measures outlined in the symmetry pattern, or they can individually seek direct accommodation with or participation in, the other international region. Yet, if countries in the Maghreb or the Levant opt for the second alternative, they do so at the cost of undermining their own potential of establishing a comprehensive Middle Eastern international region in the future.

Contrary to the "Fortress Europe" opinion which emphasizes the threat of adopting an inward-looking or closed model, an international region can therefore become a threat to the survival of other regional groupings when it is prepared, or potentially willing to concede participation in the benefits of its system to outsiders. Such a possibility presupposes that there are real rewards for interacting and cannot occur if a significant geographical distance in combination with an absence of closer political ties are not present. Hence, while Latin American countries have had little choice but to pursue the symmetry pattern to counter the EU integration process, countries in the Maghreb and the Levant, which have been tied to Europe by bonds dating back to the colonial period, could find themselves gradually being pulled more and more into the Western European regional sphere of influence.

In recent years, several efforts have been made to institutionalize dialogue between Southern Europe, the Maghreb and the Levant, as demonstrated in the CSCM, Council of the Mediterranean (CM) and Western Mediterranean Forum proposals, which are discussed at length in the next chapter. In the north, the EU member states of the Southern European subgrouping often lobby for further resources to be dedicated towards the Mediterranean. In the south, tentative attempts to activate indigenous regional dynamics have evolved unevenly. While the creation of the Arab Maghreb Union appeared to hold out the prospect that indigenous regional dynamics would increase, the isolation of Libya by the West and the crisis in Algeria have tended to side-track Maghrebi regional relations. In the Levant, the September 1993 breakthrough in Israeli-Arab relations signalled that a period of more co-operative indigenous regional dynamics could gradually gain pace. However, regional patterns of interaction in the Levant remain fragile and limited. There is little to indicate that all the countries in the Levant will participate in regional proceedings at any time soon, as evident from Syrian overtures.

A review of the patterns of interaction within and between the Southern European, Levant and Maghreb subregions clearly demonstrate that although this collection of states in the Mediterranean area are geographically proximate, they have no common agenda and do not share perceptions of where their interests lie. Some of these geographically proximate states have strong bilaterals with the Western European international region. Others are closer to the Middle Eastern international region. Links between the subregions bordering the Mediterranean and references to the notion of "Mediterraneanism" seem to largely take place to bolster the position of individual states within their own international region. The patterns of regional relations in the area do not indicate any tendencies towards an intensification in interaction between the states of the Mediterranean which would give credence to the notion of a Mediterranean international region. It therefore appears that Southern European countries see certain states in and around the Mediterranean basin as strategically and economically important and thus want them in their sphere of influence. In much the same way, countries in the Maghreb and Levant identify states in the Mediterranean area as an important security concern and want them in their sphere of influence. This basic logic explains the pattern and nature of regional dynamics operating in the Mediterranean area. Relations are intergovernmental dominant which makes it easier for the proximate states to regulate the type and extent of contact that takes place between them. Transnational ties, which by their very nature would be much more difficult to control, are limited to the energy sector and the area of Islam. As a result, rather than an area where a "new" international region is about to emerge, it is more accurate to see the Mediterranean at the end of the twentieth century as a contact point between two different international regions, Western Europe and the Middle East. References to a Mediterranean international region are better described as boundary management devices rather than boundary transcending ones. Whether the Mediterranean becomes a divide or a link between these two international regions depends on how far both sides are prepared to interact with one another (see Figure 4.2). Initiatives paying lip service to an increase in this level of interaction have multiplied in recent years and are examined in the following chapter.

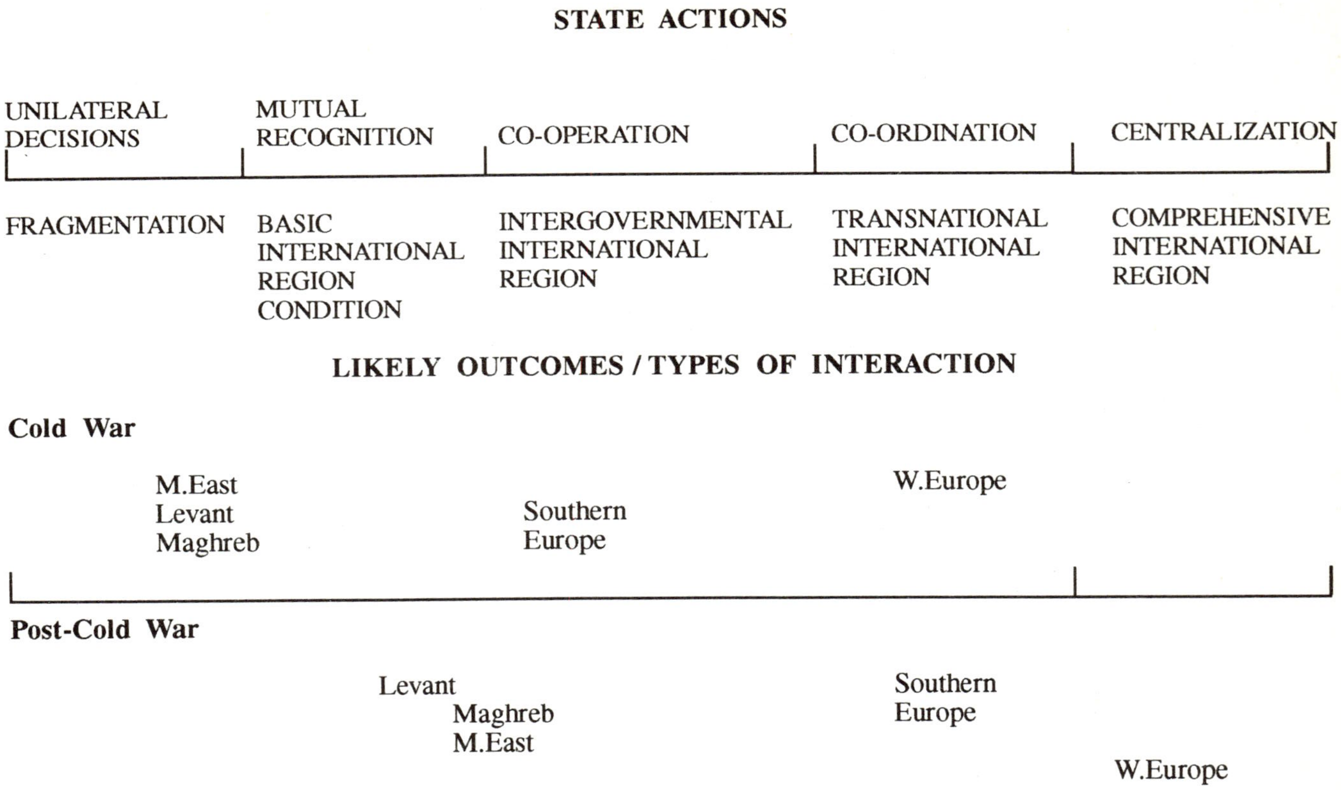

STATE ACTIONS

UNILATERAL DECISIONS	MUTUAL RECOGNITION	CO-OPERATION	CO-ORDINATION	CENTRALIZATION
FRAGMENTATION	BASIC INTERNATIONAL REGION CONDITION	INTERGOVERNMENTAL INTERNATIONAL REGION	TRANSNATIONAL INTERNATIONAL REGION	COMPREHENSIVE INTERNATIONAL REGION

LIKELY OUTCOMES / TYPES OF INTERACTION

Cold War

| M.East Levant Maghreb | | Southern Europe | W.Europe | |

Post-Cold War

| | Levant Maghreb M.East | Southern Europe | | W.Europe |

Figure 4.2 - Mediterranean Subgrouping Convergence Continuum

5 An Assessment of Trans-Mediterranean Initiatives

5.1 Trans-Mediterranean Security Initiatives

The analysis carried out in chapter three clearly shows that a Mediterranean international region existed as an historical reality. The review provided in chapter four of the contemporary situation in the Mediterranean reveals that the distinct pattern of relations within the European and Middle Eastern international regions, and more specifically within their subgroupings bordering the Mediterranean (Southern Europe, the Levant and the Maghreb), are causing fragmentation within the basin. In addition, centrifugal tendencies in most riparian states and the potential for adversity and existing diversity within and around the basin, suggests a Mediterranean sea that is not a crossroads of three continents, but a barrier where Europe, Asia, and Africa end. Nevertheless, certain protagonists of the post-Cold War system ascribe to the Mediterranean a geostrategic unity.

The aim of this chapter is to assess the prospects for regional development in the Mediterranean. First, a general overview of what determinants characterize regional processes of transformation is put forward. This is followed by an analysis of the various trans-Mediterranean security initiatives proposed since the end of the Cold War. An attempt is made to identify which modalities of regionalism these initiatives are attempting to promote across the Mediterranean. The section concludes with an assessment of how far the institutionalization processes put forward complement the co-operative and conflictual regional dynamics functioning in the contemporary Mediterranean.

It is a fundamental rule of research that concepts must be precisely defined if they are to serve any useful purpose. There is little consensus in the international region literature on the basic definitional question of what criteria best demonstrate the process of regional transformation. This study offers such a definition in an effort to reduce the "limitations, ambiguities, and inconsistencies", in existing theory (Carl Hempel, 1952: 12).

International regions are the result of changing great power configurations and variable integration and conflictual pressures. Given their fluid nature, there is nothing necessarily natural, inevitable, or permanent about them (Modelski, 1961: 150). The parameters within which the actors in any given international region must perform are constantly changing. Yet it is too simplistic to suggest that the implosion of one superpower and the limited withdrawal of the other can be regarded as conducive to the positive development of regional dynamics. The disappearance of superpower overlay removes both the constraints and the opportunities that were available to regional actors who participated in the patron-client system.

Examination of regional dynamics in the post-Cold War era indicates that there has been an increase in the intensity of relations among regional actors. It is often stated that the absence of large-scale international war and the accompanying polarization act as a catalyst to the process of regional transformation (Young, 1969: 341). Reality shows that the threat of coercion can also lead to regional integration. The range of pan-Mediterranean regional proposals since 1989 seems to confirm this trend. The ending of the superpower bipolarization along with the fear that a conflict based region could develop on religious and economic lines has led to an increase in trans-Mediterranean regional initiatives.

Second, the gradual diffusion of effective power throughout the international system, as a result of the Cold War's collapse, allows regional actors to conduct more independent external relations than was previously the case. This may even give rise to a situation where a small number of new regional power centres emerge. Developments throughout the Mediterranean demonstrate that such a diffusion of power has taken place. While the United States continues to safeguard the international sea-lanes, the vacuum left by the Soviet Union's absence has, to a certain extent, been filled by the two dominant regional organizations, namely the European Union (EU) and to a far lesser extent the Arab League.

France, Spain, and Italy, have all developed more coherent foreign policies in the Mediterranean as activities linked to projects such as the Western Mediterranean Forum, the WEU, and the CSCM demonstrate. All three have also sought to increase the EU external affairs dimension in this area. Along the southern coastline of the basin, Egypt is emerging as a regional power centre. It has already displayed the ability to influence the course of regional events by playing a direct, if not assertive role, in the Middle East peace process. Cairo can be credited with helping to convince Palestinian Liberation Organisation representatives to reach an agreement with Israel. The redistribution of power that is currently taking place in the international system therefore augurs well for the rise of "new" poles of power that will change the design of regional relations.

Third, for the first time, former colonies that became independent during the Cold War are now in a position to act more independently than before. The general rise in levels of political consciousness and the spread of active nationalism throughout the developing world are both signs of this transformation process. This is especially evident in North Africa where a wave of political fundamentalism is challenging the European colonial fabric that was left behind in the 1960s.

The former British colonies of Malta and Cyprus are also exercising their new found proactive foreign policies by seeking EU membership, which would extend the EU border into the heart of the Mediterranean. Further North, the disintegration of superpower overlay has not only seen the rapid growth in the number of independent states in what was formerly Yugoslavia, but also the emergence of new lines of conflict, as effective influence spreads to each of the regional actors, primarily Serbia.

This long-term structural trend toward international fragmentation has gathered momentum since the breakdown of the Cold War (Buzan et al., 1993: 6-8). The process is conducive to an increase in regional interaction because the lifting of external constraints and co-operation permits regional dynamics to evolve much more independently than was previously the case.

M. Haas (1970: 102) identifies two distinct determinants that demonstrate phases of regional transformation. Regional variations take place if there is a change in the number of poles of power within a particular area. This can occur in a number of situations: through the demotion of a major power in a military defeat; the elevation of one or more middle powers to major power status; the entrance or exit of a major power into or from a region; the creation of new military alliances involving major powers, and through changes in existing military alliances in which major powers collapse or membership is reshuffled.

A "reality check" of the current situation in the Mediterranean area reveals that a number of regional structural changes have taken place in the last few years. Although the Soviet Union did not suffer a military defeat, it has been demoted as a regional player after its implosion. As noted above, a few middle powers, primarily France, have demonstrated their ability to play more active roles in the region's international relations. The shift in NATO's "out-of area" philosophy, the revitalization of the Western European Union, and the disappearance of the Warsaw Pact, are all developments that illustrate the substantial structural changes that are part of the post-Cold War regional transformation process.

The second distinctive occurrence, which Haas refers to as an indicator of regional transformation, is that of power distribution (ibid.). A region can evolve when a dramatic event precipitates a trend toward tightening or loosening the power distribution. Instances of such a development include those where a previously non-aligned power becomes an ally of another major power; an alliance among major powers disintegrates and causes a schism to develop; many non-major powers become attached to or break away from bloc arrangements; or, pluralistic security communities disintegrate (ibid.).

As previously noted, a gradual diffusion of power is occurring throughout the international system of states and certain behavioural variations are already noticeable. While one non-aligned actor (the former Yugoslavia) has completely collapsed with far reaching consequences, other non-aligned states have sought refuge in various international forums to compensate for the Non-aligned Movement's relegation to the side-lines of contemporary international relations. For example, Malta's efforts to become a full member of the EU is indicative of such an approach. While paying lip service to the notion of a Mediterranean international region, Malta is systematically aligning itself with the Western European international grouping and thus reinforcing the presence of this unit in the area.

The process of European Union integration has seen a group of major powers move closer together (France and Germany) at the risk of alienating others, both within the alliance (Britain), and especially those on the Union's exterior periphery (the Maghreb and the Levant). As the EU deepens and widens further its integration project, and thus extends its southern border to the centre of the Mediterranean, it increases the risk of causing a schism between those states involved in the process of European integration and those on the periphery.

This transformation process is still at an early stage, and the indicators do not clearly show the likely future patterns of interaction, although a number of trans-Mediterranean institutional frameworks have been proposed to help forge relations in the area. An assessment of these initiatives illuminates just how close to reality such concepts are.

5.1.1 The West Mediterranean Forum: "The 5+5 Talks" (Intergovernmental Model)

The first post-Cold War trans-Mediterranean security initiative to be examined is the regional dialogue launched by the French in 1990, namely the West Mediterranean Forum. Questioning the more global approach of other institutional frameworks put forward to help regulate the dynamics of fragmentation in the area, the "5+5" process advocates a more narrow cross-regional approach to the problem. It consists of members from the western zone of the basin, where the dynamics of regional fragmentation were less advanced than those present in the eastern sector. Composed of five Southern European countries - France, Italy, Malta, Portugal and Spain - and the five Arab Maghreb Union members - Algeria, Libya, Mauritania, Morocco and Tunisia (Malta joined the grouping in 1991), the "5+5" talks were officially launched at a foreign ministerial level in Rome on 10 December 1990 (Barbé, 1992: 14-5).

This intergovernmental approach to establish a security forum in the Mediterranean consisted of a flexible structure of dialogue, consultation and co-operation. Ministerial meetings were to be held at least once a year. In the interim, working groups were set up to tackle national and transnational issues of concern such as international financial resources for development, food self sufficiency and desertification problems, debt re-scheduling, migration flows, and the preservation of cultural heritage. Although the process was supposed to embrace the concept of security in its broadest sense through the integration of political, economic, cultural, human and ecological factors, it actually concentrated most of its resources on the economic aspects of West Mediterranean relations (Ghebali, 1993: 97-9).

The Rome meeting was followed by a ministerial gathering in Algiers from 26-27 October 1991 where the issues discussed included: the creation of a Mediterranean data bank in the fields of industry and commerce, a Mediterranean financial bank, and a Eureka-type Western Mediterranean programme for science and technology (ibid.: 100-1). In an attempt to head off the instability that is generic to the Middle East, the Europeans put economic policies of co-development at the top of the Group of Ten agenda. Their main objective was to arrest population growth and dispersal, and to encourage the more efficient management of resources as a means of enhancing regional interdependence and development (Vasconçelos, 1991: 29).

Interaction of a political and social nature was severely limited in the ministerial meetings, while military issues were completely absent from the agenda. It was hoped that the strengthening of economic ties between Southern Europe and the Maghreb would help reduce their asymmetry on many other levels of interaction. Some pundits presumed that success in the "5+5" talks would result in the forum expanding to a "5+13" formula, that is, the five countries from the Maghreb plus the then twelve European Union member states and Malta. Actual developments were to prove this aspiration unrealistic (Mortimer, 1992: 41).

The third meeting scheduled to take place in Tunisia in 1992 never materialized. After a productive start, the French initiative foundered, owing not only to the problems with, and sanctions against, Libya, but also because its southern interlocateur, the Arab Maghreb Union, is itself paralysed due to the civil war in Algeria. Moreover, Algeria is perhaps only at the initial stage of a political and economic crisis which several other Arab and Islamic countries are now beginning to experience, as events in Egypt tend to demonstrate (Vasconçelos, 1994: 3).

A review of the Western Mediterranean Forum immediately reveals that one of the basic problems with this initiative is that it is attempting to place two completely different regional subgroupings on an equal footing. While intergovernmental ties across Southern Europe consist of a community of states engaged in a process of integration, the Maghreb is made up of a group of states where the patterns of interaction indicate a process of fragmentation. The lack of regional convergence at this level is made worse by the lack of transnational patterns of interaction between Southern Europe and the Maghreb to sustain the intergovernmental endeavour. The large amount of vertical trade between the two subregions is primarily conducted by the states themselves in the energy sector, with no direct benefit for the citizens of the Maghreb.

Transnational social and political interaction is more or less absent due to strict government restrictions in both these sectors. The introduction of the visa for citizens of the Maghreb (Spain 1991) and the clamp down on Arab political exiles throughout the European Union (especially France in 1994) indicates the trends in these areas (Financial Times, 13-7-94: 22). The highjacking of an Air France aircraft in Algeria at the end of 1994 also resulted in the halting of air traffic between Algeria and both France and Spain (Time International, 28-12-94: 14-21).

Only transnational interaction at a military level to some extent mirrors existing intergovernmental military alliances between certain Western countries and their Maghreb counterparts. The United States for instance, has made use of bases in Morocco under the terms of such agreements

(Vasconçelos, 1991: 31). Interaction at a transnational level mainly incorporates the purchasing of arms by political rivals in the Maghreb; construction of weapons plants of all types including chemical weapons, as alleged in Libya; and the training of terrorists and mercenaries who often operate outside the boundaries of their own state. The problem with these ties is that they promote transnational patterns of regional dynamics which are conflictual and thus detrimental to the co-operative objectives sought in the "5+5" initiative. A multiplication of such exchanges is likely to lead to the evolution of a conflict based region.

The basic weakness of the "5+5" forum is that it integrates two completely different regional subgroupings into a single institutional framework, without first correcting the lack of unity in the perceptions of the countries involved, as to what constitutes their interests. In addition, there is no scope to promote transnational patterns of exchange among the European and North African countries involved in this process. Another criticism of this grouping is that the degree of interaction that takes place is inadequate to generate patterns of interaction that would create a less asymmetrical interdependence between these two regional subgroupings. Western Mediterranean Forum deliberations focus almost entirely on political and economic issues, with little discussion of social and environmental matters, and a complete absence of a debate on military issues. As a result, references to a Euro-Maghreb partnership contain more rhetoric than substance. The countries of Southern Europe are not ready, and do not possess the means, to resolve the challenges confronting the entire Maghreb. Euro-Maghreb co-operation in general will only increase after the "five plus five" forum is again functioning. Once these intergovernmental links are again established they will have to be supplemented by the additional weighting of the other EU member states ("15+1+5" = EU member states + Malta + AMU member states). The development of such an extensive intergovernmental network of relations in an area rich in natural and human resources will improve the prospects of transnational patterns of interaction developing. Without such an incremental increase in Euro-Maghreb relations the prospects of establishing a co-operative Western Mediterranean international subregion, i.e., harmonizing and integrating relations between Western Europe and the Maghreb, will remain an elusive goal. A conflict dominant subregion is therefore more likely to emerge.

On a positive note, the "5+5" project has provided the actors involved with valuable regional convergence lessons. First, it has increased the sense of urgency to develop a preventative security strategy which can meet the new challenges of the twenty-first century, before they have evolved into outright dangers. Second, the process has also been successful in stimulating an informal exchange of views, a first step in the exercise of consensus-building and regional co-operation. Intergovernmental political and economic interaction functioned well at the initial meetings of the forum, and given the opportunity to develop, issues of a social and perhaps a military nature could have eventually been put on the agenda. Mobilizing transnational solidarity will also have to be an objective of cross-regional meetings in the future, as intergovernmental ties between the northern and southern shores of the Mediterranean are unlikely to succeed if not complemented by cross-regional dynamics at a more "grass-roots" level.

5.1.2 The Council of the Mediterranean (CM) (Intergovernmental/ Transnational Model)

Another trans-Mediterranean security blueprint, which complements the Western Mediterranean Forum's intergovernmental approach, but adds transnational patterns of interaction in the scheme of analysis, is the proposal to set up a Council of the Mediterranean (CM).

The initiative was officially presented by Malta at a symposium held in Tunisia in November 1992. Malta's Minister of Foreign Affairs, Guido De Marco, envisaged a forum that could be established on the Council of Europe model, creating the necessary facilities to involve all the parties concerned in a continuous dialogue towards the solution of problems affecting the area:

> It is my government's view that the Mediterranean Sea should not be a divide, but through the rich diversity present in all Mediterranean conjectures, can be harmonised by structures for an on-going dialogue to serve mutual interests (De Marco, 1993: 6).

The participation of all interested parties including the European Union, the Arab Maghreb Union, and the Arab League was advocated. As guide-lines for membership of the Mediterranean Council the following criteria were cited: adherence to the principles of the UN Charter, respect for the dignity of the human person and the Rule of Law, and respect for the establishment and development of representative institutions.

The structure of the CM is to consist of a Committee of Ministers and General Assembly with consultative powers, where representatives of Mediterranean states could form a Parliamentary Assembly of the

Mediterranean. This would be supported by a secretariat intended to co-ordinate the Council's activities in the political, economic, social, environmental, and cultural sectors (De Marco, 22-11-92: 5 and 14). Reaction to this proposal among riparian states was initially mixed. Among the countries bordering the Southern shores of the basin the establishment of a Conference on Security and Co-operation in the Mediterranean (CSCM) remains a priority. Several Mediterranean countries however favour the less in-depth CM initiative, and pledged their support for this initiative at the Third Conference on Mediterranean Regions organized by the Council of Europe in Taormina in April 1993 (Agence Europe, 9-4-93: 5).

Prospects that the CM can become a forum for preventive and proactive diplomacy in the short to medium term are unrealistic for a number of reasons. As a flexible security arrangement, the Council has the advantage of being able to contend with issues of both a co-operative and conflictual nature. Yet, as a forum for discussing common interests and common concerns, the CM must first develop mechanisms to nurture the notion of a Mediterranean identity in an area where such dispositions are lacking. A first step in this direction is possible by returning to a more frequent and regular level of relations between the two major civilizations that are separated by the Mediterranean sea. If the Mediterranean is not to develop into a conflict dominant region the CM must confront the trend towards isolation, which manifests itself through the rise in radical movements seeking to assert their identity in their own particular area.

Another hurdle which the Council has to resolve before it becomes a feasible enterprise is how to apply such a holistic approach to security in an area as diverse and adverse as the Mediterranean. The shift from Mediterranean-centricity to Euro-Atlantic centricity at the onset of modern history ushered in a long era of neglect, foreign domination and fragmentation in the Mediterranean. This trend became even more rapid this century as the poles of international relations shifted further out to the West and the East (Fenech, 1993: 139). The end of the Cold War has ushered in a period where East-West dynamics are changing, but there are no signs to suggest that they are returning to a situation of Mediterranean-centricity. The CM must therefore find a way of renewing international attention in an area which has historically been a centre of attraction. With the necessary diplomatic and physical resources the CM's attempt to institutionalize discourse and dialogue in the basin will become a more realistic enterprise (Rato, 1994).

Given the diversity and complexity of issues in the littoral and the weak intergovernmental and practically absent transnational patterns of interaction that extend across the Mediterranean, pan-Mediterranean dialogue is a difficult process to nurture. Initially it will have to be limited to non-sensitive security concerns. The recommendation of imitating the successful example of the 1976 Barcelona Convention for the protection of the Mediterranean Sea is a route which it might consider imitating. Discussions which focus on issues such as energy, transport, communications, and culture will permit the Council to encourage social and political intergovernmental and transnational relations. Once this alliance network materializes, the Council can approach the more sensitive areas of intergovernmental and transnational economic and military coalition building by debating questions such as economic deprivation, ethnic friction, demographic explosion, migration, environmental degradation, intolerant fundamentalism, and outright conflict.

If the CM is to serve as a vehicle for cultivating co-operative intergovernmental and transnational international region patterns of interaction, it must find a way of bringing together the factors of regional convergence and divergence which exist in the Mediterranean. Its main task will be how to marry - even if it is a marriage of convenience - the CM agenda with that of the Western European and Middle Eastern intergovernmental ones. One way forward might include the CM spurring relations in the societal sector by proposing to target specific co-operation ventures to assist in the development of poor rural and suburban areas which are identified as principal sources of migration. The CM can also seek to create networks linking professionals in the Maghreb with their fellow colleagues in Europe with the aim of motivating them to participate in the development of their own country. The advancement of such intercourse would help balance the conflictual intergovernmental rhetoric that often exists among the countries of the Mediterranean.

The CM will also eventually have to address the notion of an intergovernmental military alliance network which encompasses the Mediterranean. This is no straightforward task as evident from the difficulties being experienced in Western Europe, where the concept of setting up an interlocking security policy remains a distant goal. Succeeding to link the states in the Maghreb and the Levant with the WEU or with NATO in a framework comparable to the North Atlantic Co-operation Council (NACC), would enable the CM to request the assistance of such organizations in times of crisis (Calleya, 1994b: 35). Apart from the direct benefit that such an approach entails, promoting already existing security organizations will avoid unnecessary duplication of efforts in an area where the priority should be to invest all available monies in co-development projects.

The introduction of the CM would equip the UN, under Article 53 of the UN Charter, with a regional arrangement to enforce action under its authority: "the Security Council shall, where appropriate, utilize such regional arrangements or agencies for enforcement action under its authority" (Ghali: 1992: 6). In co-operation with the UN, the CM would thus be in a position to further enhance intergovernmental political and military ties across the Mediterranean by endorsing concepts such as conflict prevention, peace-keeping, peace-making, peace-enforcing, and peace-building throughout the basin.

Given the dominance of the regional dynamics of fragmentation which exist in the Mediterranean, it is difficult to picture the evolution of a collective security organization in the area. Before the Council can promote intergovernmental and transnational social, economic, political, environmental and military patterns of interaction across the basin it must understand the already existing regional dynamics. Introducing any of the following measures would help the CM achieve some of its objectives: disarming previously warring factions and restoring order, taking custody and possibly destroying weapon arsenals, repatriating refugees, dispatching advisory and training teams to assist security personnel, monitoring elections, advancing efforts to protect human rights, reforming governmental institutions, and promoting formal and informal processes of political participation. Executing such an extensive list of objectives will in practice not be an easy task and will be impossible unless supplemented by a United Nation's command and control structure which has yet to evolve.

The projects which the CM proposes to encourage and which were outlined earlier aim to co-ordinate trans-Mediterranean co-development partnerships and thus stimulate co-operative indigenous dynamics that are so far lacking in the basin. This initiative also seeks to enhance co-operative types of interaction that are essential if there is to be any rapprochement between the regions bordering the littoral. Such an outcome could therefore help the Mediterranean area avoid becoming a conflict based region.

5.1.3 The Conference on Security and Co-operation in the Mediterranean (CSCM) (Intergovernmental/Transnational/Comprehensive Model)

The most comprehensive of security initiatives to be proposed in the Mediterranean basin in the post-Cold War era is that of a Conference on Security and Co-operation in the Mediterranean (CSCM). The proposal finds its roots in the declarations of Italian Foreign Minister Gianni De Michelis at the Paris ministerial meeting of the Euro-Arab Dialogue in December 1989, and that of the Spanish Foreign Minister Francisco Fernandez Ordonez in his presentation at the Ottawa Open Skies Conference in February 1990 ("The CSCM non-paper", Rome, 1991; see also Ordonez, 1990: 1-8, and Martinez Report 1991).

The project was officially launched at the CSCM's Palma de Majorca meeting in September 1990 when the "Italian-Spanish Non-paper on CSCM" was circulated. Essentially it advocated a debate on security issues. Its rationale proceeded from at least three basic arguments.

First, from an historical perspective, it was argued that Europe could not neglect its southern flank without ignoring its proper roots and identity. Second, it was stressed that Europe could not be secure as long as the Mediterranean, a sphere of contact between the North and the South, remained insecure. The logic of interdependence dictated that both areas promote security linkages. The third argument put forward was largely the result of the Gulf Crisis' influence on policy-making at the time. Iraq's invasion of Kuwait in 1990 not only provided evidence of the urgent need for a crisis prevention mechanism in the Mediterranean, but also urged for the inclusion of a North-South dimension in security strategies being formulated to contend with post-Cold War problems (Ghebali, 1993: 94-5).

The Italian-Spanish CSCM proposal stated that it had to be global in its composition, including all the actors with influence and interests in the Mediterranean. It also had to be comprehensive in its mandate, and progressive in its outlook, given the complexity and diversity of the Mediterranean and its hinterlands (Barbé, 1992: 7-9). The two Southern European actors thus hoped that their proposal would act as a catalyst for more regular and extensive contacts between the two international regions circling the Mediterranean. At most, the CSCM would act as an important vehicle towards establishing a Mediterranean international society by promoting and managing interdependence between Western Europe and the Middle East. At least, the CSCM would assist in managing and regulating the Mediterranean divide if the patterns of relations within the two Western European and Middle Eastern international regions showed no signs of converging.

These objectives were to be achieved through a Mediterranean Act which would constitute the regional codification of a series of universally valid principles, as proclaimed in the United Nations Charter and the Helsinki Final Act: a security basket aimed at instilling confidence through a code of political conduct derived from the Helsinki CSCE Decalogue; an economic basket providing conditions for a more balanced economic co-development approach; and a human dimension basket, designed to the dialogue of cultures and conciliation of values (Ghebali, 1993: 96).

To date, the Italian-Spanish CSCM proposal remains on the drawing board. While the countries on the southern shore of the basin endorsed the idea, France and Portugal have been more reluctant to sign up, with the other EU member states and the United States doubtful that such an ambitious project could realistically be put into practice. Critics of the plan indicate the complexity of issues a Conference encompassing a geographical territory that spread from Iran to Mauritania would have to confront. Serious doubts have also been cast upon the applicability of a Helsinki-type process tailor made to deal with the plethora of issues that confronted Eastern and Western Europe during the Cold War (ibid.: 97).

The lack of consensus at the Palma meeting resulted in the issuing of a non-binding open-ended report. It declared that a meeting outside the CSCE process could discuss a set of generally accepted rules and principles in the fields of stability, co-operation and the human dimension in the Mediterranean when circumstances in the area permitted (Palma Report, June 1992).

Without the support of all the members of the European Union and the influential United States, the aspiration for establishing a CSCM has been put on the backburner. Even allegiance to the concept of a CSCM among Southern European states has waned since 1990. In consensus with Morocco and Algeria, France considers the widening of Mediterranean co-operation as premature, and difficult to co-ordinate with efforts such as the already troubled Western Mediterranean Forum. Portugal is also against the formation of another security institution, as it regards NATO, the WEU, the OSCE and the EU as capable enough of handling security issues in the Mediterranean. As for the two main proponents of a CSCM, both have had to reconsider their position on the question of Mediterranean security because of domestic concerns. Having experienced a tremendous political upheaval, Italy's external concerns have recently been limited to European affairs. Meanwhile, after experiencing a severe economic recession of their own and a political transition of their own, Spanish authorities have also refocused their political resources to matters closer to home. Consequently, the notion of a CSCM remains a non-starter (Latter, 1992: 13-14).

The scheme's failure to date comes as no surprise given that the objective it seeks to realize is to place two asymmetrical international regions on an equal footing and attempt to integrate them into a single institutional framework (Gaspar, 1994: 29). A comparison of the regional dynamics in both these international regions discloses the extent of their differences. Western Europe is a community of states engaged in a process of integration, while the Middle East remains a region where the dynamics of fragmentation dominates. The common boundary between these two regions separates democracy from autocracy, market economies from under-developed economies, modern societies from archaic social structures struggling to cope with the pressures of modernization. These disparities are further compounded by the contrast in demographics, the economic dependence of the Middle East on Western Europe, and the military superiority of the West (ibid., and see Tables 4.4 and 4.5).

The CSCM proposal thus attempts to institutionalize concepts associated with the notion of a comprehensive international region where such patterns of interaction do not exist. As a result it can be described as a premature initiative. The weak system of intergovernmental relations across the basin at a social, economic and military level and the absence of transnational linkages, implies that any institutional steps taken in this sector should first be concerned with nourishing transnational exchanges on an equal footing with existing intergovernmental political and environmental exchanges. Only then can an attempt to institutionalize such patterns of interaction stand a realistic chance of success.

Efforts to foster such patterns of interaction can be viewed from a spectrum which rises through a sequence of stages, each building on the one before it. For example, as limited intergovernmental political contacts already exist (at a bilateral and a multilateral level, for example the UN), measures can be introduced to strengthen these connections. This can be carried out through a multi-faceted array of instruments which include: promoting economic reconstruction in the Mediterranean area by encouraging industrial diversification programmes and liberalizing export markets so that trade and not aid is the cornerstone of relations between Europe and the Middle East; providing financial and technical assistance to prevent further environmental damage, improving the observance and legislation of human and minority rights, and introducing immigration policies that strive to maintain a social and cultural balance in the clauses concerned with immigrant entry requirements.

The establishment of governmental political relations could then be complemented by larger structures of co-ordination and consensus of an economic, cultural, and, eventually a military nature. Incremental advances at these points of contact is how economic and ecological development, and the framework for a viable cultural area, can be fostered in a location as heterogeneous as the Mediterranean. Once such an interdependent network of contacts is functioning, the CSCM can be introduced as a means of managing and monitoring relations between the two international regions encompassing the Mediterranean, and particularly, the relations of the three international region subgroupings which border the Mediterranean. In sum, the CSCM must succeed and not precede the regional dynamics it seeks to encourage. The concept of a CSCM project collapsed precisely because it was too ambitious. Its underlying 'co-operative' approach to security does not reflect the more conflictual patterns of relations which exist across the Mediterranean.

In June 1992 another CSCM-type meeting took place at the Malaga Conference which was organized by a Geneva based institution, the Inter-parliamentary Union. The main variance between the Malaga CSCM and the Italian-Spanish project is the representative criterion that the former set up: only parliamentarians from riparian Mediterranean states were entitled to the status of full participants. As a result, countries like Portugal, the United States, Russia and Britain were excluded. The Palestinians were included as 'associate participants', which led to Israel's withdrawal. Representatives from the former Yugoslavia and Algeria also failed to attend due to the domestic turmoil they were experiencing (Ghebali, 1993: 100-101).

The conference, entitled "Regional Stability", adopted a final document by consensus which was divided into three pillars. The first pillar concerning security issues suggests the sketching of a charter handling trans-Mediterranean relations; the creation of a crisis management and peaceful settlement of disputes regional centre; the ratification of a set of land based confidence-building measures; and a series of rules with the objective of transforming the Mediterranean basin into a denuclearized zone.

The second pillar focuses upon the goals of co-development and partnership, both central themes of past and present EU initiatives in the Mediterranean. The economic programme presented touched upon a broad range of issues including: the promotion of food production; trade exchanges and industrial co-operation; debt rescheduling and relief; encouraging foreign direct investment, especially to upgrade communication and transport networks; and the management and integration of tourism. Provisos were also included on the migratory movement of refugees and workers, environmental security and research, and development strategies in the field of science and technology.

The third pillar deals with the issue of human rights and is a carbon-copy of the CSCE mechanism tackling the human dimension. It advocates a policy of transparency among Mediterranean states when confronting human rights issues. It also endorses the principle that all countries in the littoral should ensure the preservation of religious, ethnic, cultural and linguistic minorities on their territory.

The Malaga document also incorporated a preamble that proved to be the most difficult phase of the negotiations. The preamble emphasized that although the conference was not mandated to devise direct solutions to conflicts in the area, its purpose was to "launch a pragmatic process of co-operation which will gradually increase in strength and coverage, generate a positive and irrepressible momentum, and facilitate the settlement of outstanding conflicts" (ibid.).

This initiative therefore largely attempts to nurture intergovernmental interaction among a much narrower number of states than the CSCM proposes. The main weakness of the Malaga proposal is that by including only riparian states it ignores the actual reality of patterns of relations operating across the Mediterranean. Inter-Mediterranean economic and political ties remain weak despite an increase in the intensity of relations within certain sectors of the basin, such as the Levant. Centrifugal tendencies continue to function as actively as ever. The fact that the Malaga initiative does not include provisions for those external actors with an interest in the area, and which the Mediterranean states are increasingly binding themselves with, such as the United States, contributes to its impracticality.

An attempt to relaunch a Mediterranean security dialogue took place at the Gymnick meeting held in Alexandria in July 1994. This initiative, launched by Egypt, brought together ten Mediterranean countries: Morocco, Algeria, Tunisia, Egypt, Turkey, Greece, Italy, France, Spain and Portugal. Malta was subsequently invited to participate at the follow up meeting. Each country met to discuss the possibility of generating a more structured process of expert consultations on set topics at both an intergovernmental and transnational level. Plans outlining what approaches can be introduced to advance such patterns of interaction were discussed at preparatory meetings on cultural issues in Rome, on political issues in Madrid, and on socio-economic issues in Cairo during 1995. The results of these deliberations were subsequently presented at a follow-up Euro-Mediterranean gathering in France.

Similar to the Western Mediterranean Forum which was launched by France in 1990, the Egyptian initiative focuses on trying to strengthen intergovernmental relations and commencing transnational contacts between the Southern European and the Maghreb subgroupings. A key difference from the "5+5" forum is that the inclusion of Cairo brings the Levant into the institutional framework for the first time, a precedent which was enhanced during the November 1995 Euro-Mediterranean Conference in Barcelona, with the inclusion of Syria, Jordan and Israel. The Mediterranean Forum meeting held in Ravello, Italy in May 1996, discussed issues of common concern in advance of the follow up "Barcellona II" conference which is scheduled to take place in the first half of 1997.

5.2 Conclusions: Prospects for Institutionalizing Regional Dynamics in the Mediterranean Area: An Attempt to Square the Circle?

Having outlined the basic elements of the various trans-Mediterranean security initiatives that have been proposed in the post-Cold war period we are now in a position to conduct an assessment of how far these structures complement the regional dynamics functioning in contemporary Mediterranean international relations.

The analysis of these security arrangements strongly indicate that countries in the Mediterranean area are trying to expand the scope of action of their present foreign policy orientation. These initiatives also act to safeguard the economic interests of those countries which have been jeopardized by the changes in the global system after the end of the Cold War. Mediterranean countries viewed the global and regional shifts, as leading to the marginalization of their interests and, in consequence, formulated various Mediterranean policies in order to stay at the centre of international relations. The addition of a Mediterranean dimension in their foreign affairs increased their policy options and diversified their external interactions.

Having examined the proposals put forward to institutionalize regional dynamics in the Mediterranean area, what are the prospects for such a process becoming a reality? The assessment in chapter four revealed that the foreign policy priorities of the riparian states are divergent and show no signs of convergence. There is nothing to indicate that an intensification of trans-Mediterranean regional dynamics is taking place. On the contrary, centrifugal tendencies are the main characteristic in these relationships. To conclude from this, it is unlikely that the riparian states focus will become centripetal, but will remain centrifugal. Despite paying lip service to issues pertaining to the Mediterranean, countries in the area are not about to abandon their foreign policy priorities which emanate from other international regions, such as Western Europe, the Middle East, or North America.

This assertion can be further understood by the fact that small states in the area have a limited ability to change the patterns of relations focus. A clear example which illustrates this point is Malta's initiative to launch a Council of the Mediterranean (CM) imitating the Council of Europe model. Although the proposal was met with widespread support, it remains a theoretical expression, as its main proponent does not have the political or economic muscle to rally potential supporters to its cause. In reality, even a concerted effort by the European Union could not begin to address all of the security challenges in the Mediterranean basin. The analysis of the patterns of relations in Southern Europe, the Maghreb and the Levant, revealed that all of the countries bordering the Mediterranean have strategic commitments outside this international waterway. Countries along the northern shore of the Mediterranean focus their attention on developments in the Western European international region. In contrast, countries in the southern and eastern sectors of the basin are primarily concerned with events in their own subregions and the Middle East as a whole. Rather than focus on Mediterranean issues, countries such as Morocco, Tunisia, Egypt, and Israel invest more of their diplomatic resources on strengthening ties with Western Europe and the United States. In short, none of the riparian states can afford to sacrifice strategic economic and military links which they have, outside this area, for the sake of a Mediterranean orientation. Thus, despite rhetoric in favour of a more active Mediterranean foreign policy, none of the countries in the area are re-ranking their external priorities in the direction of upgrading commitments in the Mediterranean.

A lack of consensus among some of the states in the area has also undermined this institutional process. For example, France was one of the first countries to reject the Italian-Spanish CSCM proposal in 1990. France was also critical of the Egyptian Mediterranean Forum Proposal, as it favoured a limited institutional security arrangement on the lines of the "five plus five" scheme which would ensure that its predominance in the area was not undermined. In similar fashion, countries along the southern shore of the basin are also critical of these initiatives which they perceive as divisive and exclusionary. Thus, while they have welcomed the CSCM and CM initiatives, they are reluctant to support more limited gatherings. The basic reason why these policies have not met with success is because they relate more to the foreign policy objectives of the states who propose them than to any serious effort to nurture a Mediterranean network of relations. For example, the Spanish-Italian CSCM project is better regarded as an attempt by both Southern European states to boost their relative positions inside the EU and, in particular, their influence in NATO, at precisely the time when post-Cold War restructuring discussions were already underway. A central weakness of all trans-Mediterranean supporters is their inability to articulate what the objectives of such a forum would be, a fundamental prerequisite, if they are to attract the resources required to lure other countries into accepting the concept of a Mediterranean regional paradigm.

Consensus is also lacking among the international organizations operating in the vicinity for such a security framework. Existing European security institutions regard the creation of a Mediterranean Forum as a duplication of efforts and a drain on already limited resources. Some analysts view a Mediterranean foreign policy orientation as a process towards diluting existing security ties. Thus an external policy dominated by Mediterranean issues would result in Southern Europe losing its favourable links with Western Europe, and the Levant and Maghreb losing their strong ties in the Middle East, Western Europe and with the United States. In the Maghreb and the Levant, Arabists and Islamists strongly reject attempts to institutionalize trans-Mediterranean security into a single framework. The former contend that a Mediterranean Forum will dilute the Arab identity and lead to the dismantling of the Arab regional system (Selim, 1994: 19). The latter view such efforts as an attempt to rally European and international support against the Islamic resurgence movement. The governments of Southern Mediterranean countries have somewhat balanced these negative reactions to pan-Mediterranean proposals by pledging limited support to such plans. In reality, such support has been forthcoming more out of fear of being marginalized by Western Europe than because of any firm belief that a trans-Mediterranean forum could function.

To date, the security initiatives put forward have been co-operative bias in an area where such regional dynamics do not exist. If such developments are to reflect a more accurate picture of reality they will have to incorporate mechanisms to contend with the dominance of conflictual dynamics. Perhaps one way of attempting to institutionalize such tendencies is to imitate conflict resolution frameworks that were favoured by the superpowers during the Cold War.

Lastly, and perhaps the most fundamental problem preventing trans-Mediterranean security initiatives from taking shape, is that the regional dynamics operating in the Mediterranean area are too asymmetrical to be put into a single institutional framework. Given the enormous socioeconomic, political and military disparities that exist between the Western European and Middle Eastern international regions which surround the Mediterranean space, it seems an impossible task to try and institutionalize so many different interest groups in one regional forum. Above all else, if the countries of the Mediterranean do not perceive themselves as sharing common goals, the political will to establish a Mediterranean forum will always be absent. In a recent study, Stubbs and Underhill say that the process of regionalism is heavily influenced by the extent to which particular groupings of geographically proximate countries have developed organizations to manage crucial aspects of their collective affairs (Stubbs and Underhill, 1994: 332). They argue that just as common historical experiences and increased socio-cultural, political, or economic links can lead to the development of regional organizations, the creation of such international organizations can also increase the linkages that bind the region together. However, what is essential is that the grouping of geographically proximate countries already share a collective identity which they can in turn build on (ibid.). This essential characteristic is lacking in the Mediterranean area and makes the task of setting up an international regional organization in the area an unrealistic objective in the short term.

Theoretically, the CSCM and the CM should stand a chance of functioning in the Mediterranean. The number of actors in the vicinity, the already existing intergovernmental links and the diversity of views is not too different from that which existed in Cold War Europe when the CSCE was first established. Yet there are two key differences which distinguish the Mediterranean space from its European hinterland.

First, the disparities which any institutional framework must bring together are much larger across the basin than they were in Europe in 1975. Although Western and Eastern European societies and states were at different stages, and perhaps on different paths of historical evolution, superpower overlay bound the two entities in ways that are absent in the post-Cold War Mediterranean. It is already clear that an Arab-Israeli peace is not going to be a panacea for the entire Arab world. The complex Arab-Israeli situation is far from being resolved, and the limited and fragile co-operative interaction taking place is still shadowed by conflictual patterns of interaction, as tension over Jerusalem and the construction of Jewish settlements illustrates (Times, 10-5-95: 8). In short, less hostile Arab-Israeli relations should not be seen as signalling the emergence of a co-operative regional system in the Levant. It seems more accurate to describe the recent phase of détente in this subregion as an acceptance by the countries involved to coexist with one another. In fact, the eastern and western poles of the Middle East seem to be under the influence of completely different patterns of regional dynamics. In the Levant, co-operative intergovernmental contacts are present after decades of conflictual relations. In the Maghreb, the opposite trend has taken grip. The Algerian crisis threatens to undermine steps towards compromise and co-operation in this subgrouping, and threatens to set back the impressive economic progress in Morocco and Tunisia (Riedel et al., 1994: 1-19). Coupled with the uncertain situation in Libya, relations in this area are ripe to provoke regional tensions which could put important energy supplies for Southern Europe at risk. A CSCM or CSCM-type forum therefore faces the monumental task of not only having to contend with bringing the Western European and Middle Eastern international regions together, but also in managing the dynamics of fragmentation which are manifesting themselves in places such as the Maghreb.

Secondly, trans-Mediterranean security initiatives lack the international political support which the CSCE process had on its side. Without the backing of the United States, the European Union, the Arab League and even the Russians, any institutional framework proposed, to help manage the international regions bordering the Mediterranean, will remain an empty shell. Western Europe on its own lacks the means to correct the socioeconomic and political disparities in the area, even more so now that it is dedicating a large proportion of its assistance to Central and Eastern Europe (Aliboni, 1992: 12-13; Financial Times, 11-11-94: 18). The United States can help make up for Europe's shortcomings but the last superpower has already made it perfectly clear that it is relinquishing its 'cheque-book diplomacy' style of doing business, for a more selective and less grandiose international position.

While intergovernmental and transnational networks of links have multiplied in the Western European international region, very different patterns have developed in the Middle East, and the dynamics operating within each international region show no signs of either converging or intensifying. Transnational flows in the Maghreb and the Levant are not only less dense than their Southern European neighbours but are practically non-existent. Thus, before one can start talking of a Mediterranean international region, the degree of interdependence between the three subregions must increase to the level indicated in the framework of analysis: "the states' pattern of co-operative or conflictual relations or interactions exhibit a particular degree of regularity and intensity to the extent that a change in their foreign policy actions have a direct influence on the policy-making of neighbouring actors" (see section 2.3).

Trans-Mediterranean exercises will only evolve if they take into consideration the existing international regional dynamics which operate around and across the Mediterranean theatre. Does this mean that a CSCM is an impossible venture to undertake, and that the notion of a common Mediterranean order is likewise an impossible aspiration?

The answer to both these questions depends on what set of objectives a CSCM hopes to realize from the start. If it is seeking to create a co-operative Mediterranean international region, then in the short to medium term it is better that such an endeavour is not undertaken, as there is nothing to indicate among the patterns of interaction between Southern Europe, the Levant and the Maghreb, that such an international region is developing. If, on the other hand, the aim of a CSCM is to develop stronger patterns of transnational and intergovernmental links, capable of creating co-operation and limiting conflict across the Mediterranean divide, then the enterprise becomes more feasible. Boundary management rather than region-creating is where the CSCM and other trans-Mediterranean security proposals have a realistic chance of succeeding.

One basic merit in creating a CSCM or CSCM-type forum, is that it approaches the challenges in the area from a global perspective. If the CSCM is to foster regional patterns of interaction that are conducive to the creation of an international region, then it must be comprehensive, universal, progressive and cumulative in its approach: comprehensive in its content - both governmental and transnational levels of regional relations of a political, military, social and economic nature should be included. Universal in its breadth of application and membership - all concerned actors (those in both the Western European and Middle Eastern international region and non-Mediterranean countries with an interest in the area such as the United States) must be permitted to participate in the proceedings of such a conference. Progressive and cumulative in the sense that advances will be made by the summation of the results. In other words, higher levels of regional interaction will be sought, only when the first designated patterns of co-operation have been attained.

Like any other system, the CSCM will have to establish its own internal balances. For example, the economic and political dimension will have to be matched by parallel efforts in the social and military sectors, if divisions separating the international regions around the Mediterranean are to be managed successfully. Politically, the West must find ways of opening channels of communication with any emerging Islamic regimes. Otherwise, the slow process of democratization in the Levant and the Maghreb may falter and give rise to an anti-Western radicalization.

Social inter-penetration is another sector which the CSCM must tackle.

Some estimates envisage as many as twenty million people in North Africa opting for emigration into Europe, where salaries are eight to ten times higher than in the South. The definitive closing of European borders, and the restrictive measures already introduced to discourage possible immigrants, may control the flow of people, at least in times of relative stability. But European countries must remember that large communities of workers originating from the Maghreb, who have made a significant contribution to the success of European industry, already exist within their territories (Joffe, 1994: 35). The perception of racist and exclusionary immigration policies towards their kin across the Mediterranean only aggravates social insecurity and could fuel the possibility of a "cold war" between Islam and the West.

Militarily, the CSCM has to find mechanisms that will enable it to regulate the crises which frequently arise in this area. This will only occur if some of the myths relating to military balances in the Mediterranean are eradicated. The perception among some European countries of an increased threat to their security from countries along the southern coastline appears to be extremely exaggerated. Statistical evidence reveals that countries in the Levant and the Maghreb are, at most, capable of defending themselves against external interventions (Greenpeace, 1992: 10-2). Since the end of the Cold War, the proliferation of conventional and non-conventional weapons in the subregions overlooking the Mediterranean sea, has not eroded the strategic superiority of Western Europe over its Middle Eastern neighbour (Gaspar, 1994: 29). If the West is truly concerned with the issue of weapons proliferation then it must stop selling military hardware to countries in this catchment area. The setting up of an international arms register as suggested by the United Nations could help in monitoring transactions in this sector.

The post-Cold War period is no longer bipolar, but not yet multipolar. East-West tensions are gradually diminishing but are still complemented by North-South strains. This trend gives rise to the following paradox: while strategic global considerations have fostered a resurgence of co-operative efforts, regional developments seem to forebode conflict. Regional security dynamics in developing international regions such as the Middle East, are evolving against a background in which more developed international regions such as Western Europe, are also entering a period of transformation. A shift from centralized bipolar intervention to a more decentralized international security system is occurring, in which indigenous patterns of regional security play a far more dominant role. In some areas one can envisage the emergence of "a group of states whose primary security concerns link closely together and that, consequently, their national securities cannot realistically be considered apart from one another" (Buzan, 1991a: 190). While such patterns of relations can be identified within the Western European and Middle Eastern international regions, no such trend exists between the two international regions encompassing the Mediterranean.

Although the end of the Cold War has seen the curtain fall on a security system based essentially on East-West confrontation, it is too early to discern with sufficient clarity the emerging co-ordinates of the future security system. Unlike the European continent, where the fall of the Berlin Wall ushered in a period of reconciliation, the Mediterranean has remained a frontier area. The basin remains a point of convergence and intersection of different political and ideological systems with major economic, social, and political disparities prevailing between its shorelines. Western European and Middle Eastern international regional divisions and conflict continue to be the hallmark of Mediterranean interchange.

The post-Cold War era has seen the process of polarization between the northern and southern shores of the Mediterranean continue unabated (European Parliament Report, 1993: 4). This structural development arrests the notion of establishing an institutional framework geared to manage co-operative patterns of relations. CSCM-type initiatives which attempt to navigate regional dynamics operating in the "Euro-Med-Arab" space have a more realistic chance of succeeding if they are constructed as conflict prevention forums. Such pan-Mediterranean projects are therefore better seen as an attempt to nurture momentum towards intergovernmental and transnational patterns of interaction during a period of fast, and largely unpredictable, transformations, at both the regional and global level of relations. A flexible approach in such initiatives is conducive to managing the numerous security challenges in the Mediterranean area (see Table 5.1).

Security Initiative	Area	Dynamic
Western Mediterranean Forum	Western Mediterranean International Subregion	Intergovernmental
Council of the Mediterranean	Mediterranean International Region	IG/Transnational
CSCM	Mediterranean International Region	Comprehensive

Table 5.1 - Trans-Mediterranean Security Initiatives and their Corresponding Dynamics of Interaction

6 The Role of Great Powers

6.1 Introduction

The purpose of chapters six and seven is to focus attention on the underexplored area of the impact external actors have on regional development. In the Mediterranean area external actors of all types such as the United States and the European Union, have little difficulty regarding the Mediterranean as a single region (Cantori, 1994: 24). Moreover, references to a Mediterranean region have multiplied in recent years (International Herald Tribune, 20-10-94: 1). Yet, a review of regional patterns of interaction in the Mediterranean area actually reveals that distinct subregional relations are evolving with little or no signs of convergence between them. This chapter attempts to identify why certain external actors pay lip service to the notion of a Mediterranean international region when regional dynamics to support such a concept clearly do not exist. A constant theme throughout this chapter is that it is essential to include an analysis of the impact external actors have on regional development if the process of regional transformation is to be better understood. I therefore investigate the reasons behind nonregional powers' involvement in the internal development of international regions and also explore the extent to which external actors can influence the basic patterns of regional relations. The structure of this chapter is as follows: after a brief historical assessment of external influence on regional patterns of interaction in the latter part of the twentieth century, section 6.2 provides a summary of the track record of the United States in the Mediterranean. This is supplemented by an examination of France's actions in the area. Although a Mediterranean great power, France is included in this analysis because its historical and contemporary continental and global links qualify it as a distinctive power within this area. A comparative analysis of the roles of the United States and France in the Mediterranean reveals the different foreign policy considerations non-regional and regional great powers confront when interacting with international regions.

In summary, this chapter provides a better understanding of what impact intrusive bilateral action has on regional dynamics in the Mediterranean.

6.2 The Role of External Powers

In order to achieve a thorough understanding of international region dynamics, I underlined the importance of assessing the impact external powers have on the process of regional transformation in my definition of an international region (section 2.3). This chapter clarifies what role extraregional actors play in the evolution of international regions.

External activities in an international region can occur for numerous reasons. Outside major powers have often become involved in a region because of their rivalry with one another. When rivalry between major external powers has been intense, they have typically been more ready to become involved in regional politics. Conversely, when their rivalry has been more co-operative, they have appeared less concerned about regional relations. For example, throughout the Cold War, the superpowers became entangled in regional affairs as a means of containing each other's spheres of influence. Such regional entanglements often left the major powers hostage to the relationships between the states of each region. The diminution of rivalry between the two superpowers since the end of the Cold War is already manifesting itself in regional patterns of relations. The disappearance of the Soviet Union has allowed the United States to become much more selective in its foreign policy areas of engagement.

Circumstances which have not been affected by this turn of events are those where external powers have been engaged in international regions to pursue specific interests. External interaction in the Persian Gulf is an example of such intrusive behaviour. External powers are attracted to the area because of their sense of dependence on the reliable flow of oil.

External powers can also become involved in international regions to act as "balancers of power". Extraregional powers can be invited in by any one of the regional actors seeking assistance to help preserve or consolidate their position within the international region. The more intense the regional adversity, the more urgently external assistance will be sought. In reality, major great powers could bully their way into regional relations, but their involvement is greatly facilitated by the presence of regional rivalries. Wriggins classifies external involvement in international regions under two headings: "the pull factor" and the "push factor". The first dynamic operates when regional actors issue invitations to non-regional powers. The second dynamic operates when competition among non-regional powers leads them to seek client-states to help bolster their position (Wriggins et al., 1992: 11).

A number of factors can contribute to an increase in acts of intervention by external powers in a particular area. Firstly, international systems encompassing large number of states which endure high levels of internal instability are likely to have a high incidence of intervention. External powers are likely to intervene for two reasons: to gain a foothold in the area; to prevent any one actor from dominating the region.

Secondly, regional systems dominated by ideological divisions and competition are prone to military interference. States within international regions seeking to become regional power centres will interfere in the affairs of their neighbours to upset the balance of power in their favour.

A third systemic factor stimulating intervention is asymmetry in the distribution of power. More or less equal states have the capacity to resist each other's attempts to intervene in their internal affairs. In such systems where the distribution of power is equally shared, the incidence of intervention will be low. By contrast, systems in which power is unevenly distributed will be intervention-prone.

The intrusive sector consists of those external actors who are politically engaged in a direct manner in the international relations of a region. An external actor becomes a significant player in an international region or subregion when its actions affect the balance of power within this grouping. A country can interact with a region on a one-to-one basis at a bilateral level, or as a member of a larger organization through multilateral diplomacy (see chapter seven). In contrast to the specialized nature of bilateral arrangements and ad hoc coalitions, multilateralism is a more demanding type of diplomacy that involves co-ordinating national policies among at least three or more great powers (Kegley, Jr. and Raymond, 1994: 214). One approach is 'state-centric' and the other is 'multi-centric' in nature. The two are mainly separated by the principle of sovereignty, or rather, the adherence given to this principle by political actors (Rosenau, 1990: 9-10).

As the last superpower, the United States qualifies as one of the principal intrusive actors at both levels of interaction. It projects enough political, economic, and military authority in the Mediterranean area to influence the international relations of the Mediterranean through its strong network of bilateral contacts in the area. These include relations with Israel, Egypt,

Morocco, Portugal and Italy. The United States' participation in international organizations such as NATO, the OSCE, the UN, the World Bank and the IMF allows it to penetrate the Mediterranean area from a multilateral perspective and enables it to influence the pattern of sub-regional dynamics in Southern Europe, the Levant, and the Maghreb in a more covert manner.

France is included in this section so that a comparative analysis can be conducted between regional and non-regional great powers in the Mediterranean. Although a Mediterranean great power, France's continental and global interests qualify it as a distinctive actor. Historically and in contemporary international relations Paris has also demonstrated the ability to affect Mediterranean politics through bilateral and multilateral channels. Its influence in the Mediterranean basin is concentrated in the western zone, with a continuation of bilateral links with former French colonial outposts in the Maghreb. In the post-Cold War era, Paris has also continued to follow a proactive foreign policy. This included the launching of the multilateral West Mediterranean security initiative, otherwise known as the 'five plus five' talks.

The United States and France both maintain strong diplomatic, intergovernmental and transnational links through state visits and commercial contacts with influential elements in the basin. For example, Israel, and more recently Egypt, have enjoyed extensive military and political support from the United States. The multiplicity of economic, cultural, and military ties existing between these regional power centres and the United States reflects the importance attached to Israel and Egypt as a stabilizing and friendly influence in the Middle East.

Very much in the same vein, France has opted to cultivate its intergovernmental and transnational ties with countries in the Maghreb, having identified Mediterranean regional powers that espouse ideological doctrines and philosophies that are popular among French-Arab citizens who number close to two million within France.

Despite these active great power roles in the Mediterranean, there are few examples in recent years of resistance being applied by great powers to restrain expansionist policies of regional actors in the Mediterranean. The United States became involved in regional affairs when it bombed Libya in 1986, but this incident has been the exception rather than the rule. For example, France has to date avoided direct military intervention in Algeria. External actors activities tend to be more concerned with the human rights and terrorist policies of these countries than their regional power aspirations.

6.2.1 The Role of an External Great Power: the United States

Throughout the Cold War, American interest in the Mediterranean area was largely shaped by the mutual rivalry it shared with the Soviet Union. In February 1947 Britain informed the U.S. that it was no longer able to guarantee the independence of Greece and Turkey. Faced with the choice of filling the vacuum of the British withdrawal or permitting the area to enter the Soviet orbit, the U.S. chose to protect the area through the Truman Doctrine of March 1947. This development represented the formal aspect of an American commitment to the Mediterranean. In a limited way, the American presence was reminiscent of that of Britain in previous centuries: it provided the U.S. with a foothold for achieving desired ends elsewhere, namely in relation to continental Europe and the Middle East region.

The British structure of authority coupled with U.S. economic and military resources provided the cornerstone of U.S. policy that was to prevent any Soviet hegemonic threat to Europe or Africa. At the height of its power in the Mediterranean basin, the U.S. had the following facilities it could utilize in a Mediterranean crisis: military bases in Morocco, Libya, Saudi Arabia, Portugal, Spain, Italy, Turkey and Greece. It could also call on British bases in Gibraltar, Malta, Cyprus, and Aden.

Two reasons help to explain the rationale behind American containment policy: to counter Soviet efforts in the eastern Mediterranean and in the Middle East, and to counterbalance the actual projection of Soviet military power into continental Europe. The U.S. made use of Italy in the West and Greece and Turkey in the East to realize this policy of containment. At no time were the internal affairs of the Mediterranean countries considered as important in themselves. Washington was strictly interested in maintaining a string of bases from which it could monitor any regional patterns of interaction that could alter the balance of power against it. The Mediterranean was therefore a strategic unit from which the U.S. could project its foreign policy goals.

The raison d'être of American involvement in the Mediterranean during the Cold War can be summed up in order of priority as follows:

* ensuring the free flow of oil to the Western world, particularly Western Europe;
* guaranteeing free access to the sea-lines of communication;
* enhancing the political and military cohesion of NATO and defending its continental Europe against Soviet pressure;
* countering Soviet attempts to gain influence throughout the Middle East, particularly the Persian Gulf, but also the Levant and North Africa.

The U.S. was thus perceived as the guardian of Mediterranean stability by its allies, and the custodian of the status quo by its enemies. Although the U.S. was under constant pressure to monitor Soviet actions, challenges to America's position in the Mediterranean tended to come from two other sources independent of Moscow. First, militant Arab nationalism which was a reflection of the Arab-Israeli conflict and, in the eighties, manifested itself in international terrorism. Second, unsettling domestic trends in some of its NATO allies, particularly the tense relations between Greece and Turkey.

Washington's foreign policy objectives in the Middle East have been both global and regional in nature (Atherson, Jr., 1984: 1194-1209; see also Spiegal, 1985). Up to the beginning of the 1970s, East-West competition dictated that the support of regional actors be one of America's main concerns in this area. This was especially the case when both Egypt and Syria turned to the Soviet bloc for arms (1955), thus permitting Moscow to gain its first foothold in the Arab world.

American intervention in the Suez affair helped it to shed its image as a new colonial power among some Arab states. But the Iraqi revolution in 1958, the 1967 Six Day War and 1973 Yom Kippur War quickly undermined Washington's attempts to become an effective mediator in the Arab-Israeli conflict. Although the conflict itself did not threaten Western European security directly, apart from the economic panic it caused after the price hikes in 1973, the threat of a Middle East apocalypse has often been a source of friction between the U.S. and Western Europe (Campbell, 1980: 165-86). More dependent on Middle Eastern oil supplies, more vulnerable to the threat of terrorism and given their deeper historical links with this area, European countries have been more inclined to be sympathetic to Arab demands than their American counterparts. For example, Spain, Greece and Turkey joined members of the European Community in 1973 in refusing Washington access to their bases and facilities to support Israel.

The presence of the U.S. Sixth Fleet demonstrates Washington's significant interest in the Mediterranean. During the Cold War the structure of the Fleet consisted of two carriers and approximately fifty surface ships. The rationale for the Sixth Fleet was traditionally based upon East-West considerations, that is, to bolster NATO's southern flank and to participate in U.S. nuclear deterrence. On two occasions, the Arab-Israeli wars of 1967 and 1973, the Americans and Soviets engaged in a fierce balance of power struggle with their respective clients in the area, and in the latter incident the U.S. was even put on strategic nuclear alert (Weinland, 1979: 7-53).

Despite recent cutback announcements in deployable carrier battle groups, the Fleet continues to fulfil both military and political roles. Most Mediterranean states are in favour of it remaining in the littoral as an insurance against potential forces of instability (Janssen Lok, 1991: 747; Latter, 1992: 10).

Technological developments have however tended to reduce the role of sea power. Changing aircraft and missile technology coupled with advancement in lift capabilities and developments in projecting power have resulted in the situation where land-based systems have become far more dominant in the sea combat environment. This is particularly the case in the land-locked Mediterranean. It is primarily in the 'choke' points of the basin (the Straits of Gibraltar, Suez, the Channel of Sicily) which are obligatory points of passage, that maritime power in the form of submarines, remain a dominant force. Submarines retain their comparative advantage due to the high thermal gradients, the elevated salinity of the sea, the uneven conformation of the seabed, and the heavy traffic (Cremasco, 1979: 13-23).

On the northern shore of the Mediterranean, American policies have been directed towards preserving the status quo throughout NATO's southern flank. One challenge it had to face was the rise of Eurocommunism among Southern European political forces and their commitment to unilateral disarmament. Another difficulty has been to maintain coherent co-operative relations with both Turkey and Greece, both members of NATO. In both situations, Washington employed carrot and stick tactics to preserve its base rights. Intensive diplomatic negotiations combined with an increase in financial assistance enabled Washington to reach agreements with Socialist governments in both Spain and Greece. The threat of withdrawing assistance to both Athens and Ankara has also permitted the U.S. Sixth Fleet home-porting rights for most of the last five decades - Greece withdrew this right between 1974 and 1980 after the Turkish intervention in Cyprus.

More recently, Europeans readily make a distinction between NATO and non-NATO contingencies, and are more likely to offer support to American forces in the former case. Plenty of ambiguity surrounds European reservations on this issue ranging from a genuine desire to maintain their sovereign prerogatives on the one hand to a quest to obtain the highest price possible through bargaining techniques for access to bases on the other.

American foreign policy in the 1990s is evolving from the concept of the New World Order, outlined in the 1990 National Security Strategy of the U.S. document by former president, George Bush (Anderson and Fenech, 1994: 15; see also Dismukes and Hayes, 1991). United States national interests in the area are geopolitical in nature:

* to assure security of access to oil reserves from   the Persian/Arabian Gulf;
* to maintain strategic and political access to Israel;
* to nurture American-Arab relations in the area, for example the strong ties that exist with Egypt and Saudi Arabia.

The U.S.'s significant role in the Kuwaiti-Iraqi crisis and its defence agreements with Tunisia and Morocco, have enhanced its reputation among most Arab oil states (Joffe, 1994: 24-5). The focus of American interests has gradually shifted to the south and east of the basin, where Washington has intensified its bilateral contacts with countries such as Turkey, Israel and Egypt, and through NATO's AFSOUTH command and control (Latter, 1992: 9; see also Lesser, 1992: 1-10).

The main challenge confronting the U.S. in the Mediterranean now is political in nature: how to justify domestically the presence of such a formidable force now that the Soviet threat has disappeared. Maintaining a network of military alliances and bases in the basin is further complicated by the asymmetrical patterns of interaction which exist between the two international regions bordering the Mediterranean, namely Western Europe and the Middle East (see Holmes, 1995: 213-35).

In the past, the U.S. has played a double role in the littoral as the strategic guarantor and crisis manager of disputes. The more erratic post-Cold War regional dynamics operating in the Mediterranean will make both assignments more difficult to accomplish in the future. Keeping the sea-lanes open for the free flow of oil to pass through will remain the Americans crucial goal. But Washington has already signalled that in the post-Cold War world it expects the Europeans to share in this burden as demonstrated in its arms-length approach to the Yugoslav and Algerian crises.

Although its global commitments and aspirations dictate that the U.S. maintain its presence in the Mediterranean area, it is still a relevant exercise to speculate what would happen if Washington were to withdraw its forces from the Mediterranean, much like the British did in the Indian Ocean and the Gulf during the first half of the twentieth century. An American exit from the Mediterranean would immediately result in a power vacuum in the area. Such a development would enable regional power centres such as Israel, Turkey and Libya to conduct more autonomous foreign policies than currently the case. An American withdrawal could also see bilateral types of external intervention in regional affairs make way for multilateral types of intervention. As a result, international organizations would become the dominant nonregional actors in Mediterranean international relations. International organizations such as the European Union are therefore likely to assume a more active external policy in the Mediterranean in an effort to consolidate their sphere of influence. This would have a profound impact on both intergovernmental and transnational patterns of relations, as countries in the Mediterranean seek to come to terms with the different type of intrusive sector operating in their area. The multilateral nature of international organizations could make it easier for riparian states to accept their involvement in regional politics. Conversely, riparian states could regard an increase in international organization activity as a covert attempt by external powers to control their regional relations.

In theory, one may argue that a U.S. exit from the Mediterranean would push countries in the area into harmonizing their foreign policies and adopting a common Mediterranean identity. However, centrifugal tendencies in most riparian states indicate that such an outcome is highly unlikely. Moreover, a total withdrawal would make it practically impossible to craft a credible regional security structure (Snyder, 1993: 109-10). The Gulf War illustrated that Southern European countries are prepared to coalesce in a crisis if the United States is willing to lead such a coalition of forces. Without U.S. supervision, and especially military assistance, it is highly unlikely that the countries in the Mediterranean could muster the necessary military and political will to act effectively in the field of conflict resolution. Again, the Balkan crisis is indicative of this point.

An American withdrawal is most likely to result in a temporary external power void. This emptiness would probably soon be replaced by the European Union and its embryonic defence pillar, the West European Union.

6.2.2 The Role of an Internal Great Power: France

At the start of the twentieth century France regarded itself as the 'new Rome' having restored its former ties that existed between North Africa and Europe (Obdeijn, 1994: 2). This perception was in fact an illusion. The Second World War revealed the fragility of the Franco-Maghreb bond. Under the guidance of their supreme tactician, General Rommel, the Germans were able to control supply lines stretching across North Africa. France became an ineffective Mediterranean power once Paris had succumbed to the German invasion. It was left to the British to keep the sea-lines of communication open in the Mediterranean, a task they only narrowly succeeded in. When the Second

World War finally came to an end the French found that the U.S. was now the dominant external power in the Mediterranean. The Truman doctrine in 1947, the arrival of the Sixth Fleet in 1948, and the creation of the Atlantic Alliance in 1949 had all strengthened Washington's position in the basin.

France's impressive colonization process - which included the settling of two million Europeans in the Maghreb and the introduction of economic, cultural and education policies which were geared to the assimilation of the local population with the mother country - could not remove indigenous social movements. Even the annexation of Algeria and its administration from Paris, could not prevent the decolonization movement from gaining momentum in the area. The Suez fiasco of 1956 and the Algerian war until 1962 underlined the erosion of France's former influence in North Africa.

Tunisia and Morocco obtained their independence in 1956. After experiencing a bloody liberation struggle, Algeria achieved independence in 1962. Attempts by France to maintain institutional ties with the three North African countries, first through the Union Francaise and later through a system of *independence dans l'interdependence* both failed, as North African leaders perceived such models as a veiled continuation of their dependence on the French state (ibid.).

France's influence in the area was somewhat diminished, but economic and cultural ties remained strong and U.S. inroads in Morocco and Tunisia never replaced France's dominant role in the Maghreb. Despite their political independence, the Francophone African states remained within France's sphere of influence by virtue of their historical ties and geographical propinquity (Martin, 1989: 104).

This Franco-African version of the Monroe doctrine continued throughout the various political regimes of the Fifth Republic, from 1958 until 1988. A central development that has made Paris reassess its role in the Maghreb is the European Union process (formerly European Community) which evolved from the Treaty of Rome in 1957. Initially, European integration did not seem to have too many repercussions on relations with North Africa. Yet, the admittance of Spain and Portugal into the EC in 1986 obliged France and the other member states to limit the import of a wide selection of North African products into the EC market.

In concert with Germany and its other EU counterparts, France has endeavoured to play a leading role in a new "Euro-Arab" dialogue which was launched through the Venice declaration of June 1980 (Carle, 1992: 43). This multilateral approach to security challenges in the Mediterranean area is further evidenced by the importance France attaches to developments in NATO, the

OSCE and the WEU. Anti-NATO rhetoric has receded since the end of the Cold War. France has also succeeded in convincing other Atlantic Alliance members that NATO must extend its influence in the Mediterranean. At the start of 1995, NATO announced it was prepared to commence a dialogue process with five countries in the Mediterranean, namely Mauritania, Morocco, Tunisia, Egypt, and Israel. France's activities within NATO have however been balanced by a desire to see an increase in European co-operation in security and defence matters. The WEU, under the auspices of the EU, is regarded as the best vehicle to effectively carry out such a policy.

The success achieved by the Maghreb countries, in the form of political independence, has not been matched by similar developments in the cultural or economic sectors where dependence upon France remains largely intact. Some observers even describe French policy in the area as neocolonial in nature, in the sense that it is designed to preserve a status quo that is clearly favourable to the western world in general, and France in particular (Martin, 1989: 101).

The main bulk of exports from Morocco, Algeria, and Tunisia continue to go to their former colonial ruler. Thirty years of independence has not resulted in diversification of the Maghreb economies whose dependency on agricultural produce such as olive oil and wine, and natural resources such as phosphate, oil, and gas, continues.

Franco-Maghreb trade relations remain fundamentally asymmetrical. Over fifty per cent of the Maghreb's trade is conducted with the EU, while horizontal trade between the Maghreb countries is practically non-existent, remaining at a stagnant three per cent level. In comparison, internal trade among EU member states has climbed to sixty per cent in the 1990s (Spencer, 1993: 49-54; see also Blin and Parisot, 1992).

Cultural ties remain somewhat intact through the neocolonial type system set up by the French. A de facto alliance between the French and African ruling elites, many of whom had studied in France or had received their education in the French language, often end up defending each other's interests at the expense of the indigenous peoples. The presence of 650,000 Moroccans, 900,000 Algerians, and 250,000 Tunisians in France also serves as a source of active contact between the Maghreb and France (Khader, 1991: 43).

Military risks are often perceived as having increased in the last few years. A direct result of weapons proliferation along the southern rim of the Mediterranean, and especially in the Levant, was the French global arms race control plan prepared for the July 1991 meeting of the five major arms control exporters in Paris. France emphasized that proliferation control depends not only on export restrictions in the North, but also on regional confidence-building and arms control in the Middle East - notably, but not only, in the Levant, where an estimated fifty per cent of global arms sales take place.

The growing debate of confidence-building in the Mediterranean has elicited growing interest in France since the end of the Cold War. In order to forestall the non-military issues in the Mediterranean region and to maximize the role of the EU as a pole of stability in the area, France proposed and launched co-operation among the "Group of 5+5", (this experiment is discussed at length in chapter five, section 5.1.1).

France is, however, in disagreement with Spain and Italy on the need for a Conference on Security and Co-operation in the Mediterranean (CSCM), tailored along the lines of the OSCE, and encompassing the entire littoral of the Mediterranean. In consensus with the countries of the Maghreb, France considers this widening of Mediterranean co-operation 'premature', and difficult to combine with a deepening of co-operation in the Western Mediterranean (Carle, 1992: 48).

The rethinking of France's security policy after the sweeping changes of the last few years is proving to be an arduous task with, as yet, uncertain outcomes. As a result of its geographical, historical and cultural realities, France is, has been, and will remain a continental European, an Atlantic, and a Mediterranean power. Above all, the history of France's relations with the Maghreb countries (its colonial past and economic and linguistic ties and the place of North African immigrants in the French community and economy), have made Algeria, Morocco, and Tunisia partners, whose "special relationship" with Paris, require constant and close attention.

In France, as elsewhere, a largely unresolved debate concerns the types of security contingencies that face Southern Europe. Many of those risks are expected to be non-military, pertaining to demographic shifts, and immigration, as well as environmental issues. France perceives the most acute challenges as originating in the Maghreb (ibid.: 50). France has played a direct security role in North Africa by intervening in a number of crises. Paris sent military aid in the form of three thousand troops to N'Djamena in 1983 to man a cease-fire line along the sixteenth parallel, right across central Chad. Continuous military provisions to the Hissan Habre's regime from France and the U.S. enabled his forces to counter the Libyan-backed onslaught of Oueddeye's forces in 1987 and led the parties concerned to take the Aozou Strip dispute to the International Court of Justice at the Hague (Joffe, 1994: 28).

The Algerian crisis has also led France to strengthen its military support for Morocco and Tunisia, supplying most of their military equipment and training in the last five years (ibid.: 25). Some French strategists have argued that the acquisition of missile technology by Algeria could directly threaten France if the former were to become an Islamic state (ibid.).

The communiqué issued after the Group of Seven summit in Naples in July 1994 represented a compromise between the positions of France on the one hand, and the United States and Italy on the other. While both Paris and Washington agree that a form of dialogue between political factions in Algeria is the best way forward, French leaders feel constrained by domestic considerations. If a situation resulted in which the fundamentalists either share power with the military or seize power outrightly, this would spur the political fortunes of far-right political factions operating in France.

French officials also fear the domino effect of an Islamic Salvation Front (FIS) take-over in Algeria. Some argue that Tunisia would fall in a matter of months and that both Morocco and Egypt would quickly be threatened. French qualified support for the current Algerian leadership is based on the belief that an FIS victory would spell trouble among the large Algerian and broader North African community in France, which estimates put at two million. With unemployment levels already at fifty per cent in some areas, the fear is that an FIS take-over would unleash an influx of refugees towards France and thus exacerbate the already chronic unemployment situation.

In what was widely regarded as an attempt to correct the favouritism shown towards Central and Eastern Europe by the Germans when they held the EU presidency in 1994, one of France's objectives during its presidency in 1995 was to organize a forum for co-operation involving all Mediterranean states (Statement by Alain Juppe at the Forty-Ninth Session of the United Nations General Assembly September 1994).

France's EU agenda reflects the significance Paris attaches to the Mediterranean dimension of its foreign policy. Its geopolitical interests in the Maghreb are determined by three principal factors: the size and degree of its economic investments, the number of French residents in the area, and the nature of the ties that exist between Paris and the national ruling elites (Martin, 1989: 115). This cluster of factors coupled with the role of supply routes through the Mediterranean means the area in question will continue to be a permanent factor in French defence policy. But France is increasingly adopting the view that the politico-economic strategies required to contend with the non-military issues in the Mediterranean can only be worked out at the EU level.

The champion of European interests in the Mediterranean is therefore France. Together with Britain and the U.S., France was a major arms exporter to the area throughout the 1950s. As long as France was beset with a nationalist revolt in Algeria, close relations with Israel were favoured. A de facto alliance was forged which guaranteed Israel a continuous arms supply to counterbalance the flow of Soviet arms to Syria and Egypt. When the Algerian crisis ended, France could again reassert her historic interests in the Arab world (Brecher, 1969: 133). This shift in policy developed gradually. A rival of American global power since Suez, the French resented American support for the decolonization process of North Africa. America's alignment with Israel was controversially balanced by French assistance to the Arabs, particularly its former colonies during and after the 1967 Arab-Israeli war. France's *volte face* against Israel is regarded as one of Tel Aviv's severest set-backs since the creation of Israel in 1948 (ibid.: 134).

In the 1990s the Maghreb and the rest of the 'Franco-African community' is the only area where France retains enough power and influence to support its claim to middle power status in the international system. But a series of strategic set-backs in Syria in 1946, Vietnam in 1954, Suez in 1956, and Algeria in 1962, convinced Paris that multilateral diplomacy should be employed more effectively when confronting international crises. The Arab-Israeli situation illustrates this philosophy. In tandem with the Americans, the French and their European partners advocate a multilateral economic and political approach to resolving the Middle East conflict. Franco-American co-operation can be identified as one factor contributing to the peace process momentum since the end of the Cold War. It is the U.S. and its European allies, together with the Russians, who sponsored the current round of Arab-Israeli peace negotiations which commenced in earnest in Madrid in 1991.

America's desire to adopt a more evenly balanced post-Cold War foreign policy, through a burden-sharing formula with its allies, can add further mileage to Franco-American co-operation in the future. This assumption fits in with the current U.S. stance that its primary strategic interests in the Mediterranean lie in the eastern sector of the basin, namely the Levant and the Persian Gulf areas (Financial Times, 15-7-94: 8). By extension, one can forecast that Washington could enter into a division of labour agreement with Paris granting the French and their Southern European neighbours supervision rights in the western Mediterranean (Atlantic News, 26-11-93: 2).

When looked at from a riparian perspective, Franco-American action in the Mediterranean could be reviewed from two different perspectives. Paris and Washington's involvement could be regarded as a form of insurance against hegemonic ambitions that regional actors might harbour. On the other hand, advocates of a Mediterranean international region are likely to see Franco-American co-operation in the area as an obstacle to an intensification of regional development across the Mediterranean as both great powers seek to manage, and by extension contain, regional dynamics within their respective spheres of influence.

6.3 Conclusions

This chapter is concerned with assessing the impact of great powers' influence on regional development in the Mediterranean area. In the post-Cold War era great powers have three policies from which to select, as they attempt to implement a stable form of multipolarity. They can act unilaterally, they can advance bilateral relations, or they can engage in multilateral collaboration. Most great powers employ a mixture of the three in their foreign policies.

Unilateral great power action has decreased in the 1990s. Two reasons help to explain this trend. First, the escalation in the cost of mounting foreign campaigns makes it preferable to seek other great powers to share in the expense. Second, any great power acting unilaterally in a far away region runs the risk of being isolated by the other main actors in the system. Unilateral action is thus usually reserved for situations where a great power is under direct threat of attack.

Contact continues at a bilateral level as great powers extend their alliance network on a one-to-one basis. However, in volatile areas such as the Levant, it seems impossible to develop comprehensive bilateral relations with all the countries in the region, as the position of the U.S. illustrates. As a result, great powers have become much more selective in their bilateral candidates, exempting those relations that will not further their self interests in a specific international region or subregion.

An assessment of the types of interaction that occur between the littoral states and their external allies reveals that interaction is political and economic dominant. Intergovernmental forms of co-operation are mainly concerned with these types of links although some progress has also been registered in the environmental field as evidenced by the Rio Summit and the Barcelona Convention in the Mediterranean. Cultural and military ties remain ad hoc. No progress has been made to establish regular cultural contacts between Western Europe or the Middle East (particularly between the Southern

European, Maghreb and Levant subregions) or with the external actors in the area. For example, no concerted European effort has been made to encourage academic exchange visits from the Maghreb to Western Europe, perhaps on the lines of the EURASMES programme, which is necessary if out-of-date perceptions and information inherited from a colonial past are to be overcome. As a result, external actors are often forced to base their assumptions on regional configurations without a first hand understanding of the social fabric or cultural dynamics at work within these regions.

An attempt has been made to detect shifts in intrusive approaches toward regional groupings since the end of the Cold War. Wriggins accurately describes the regional dynamics at work between external and internal actors during the Cold War. The process of seeking external support explains the facility with which the superpowers could intervene in regional affairs. The "pull" of the invitations from the regional states was supplemented by the "push" of the external powers, either pursuing self interests or acting out their mutual rivalry (Wriggins et al., 1992: 293). Although Wriggins' analysis is concerned with regional relations in South Asia, his summary on the regional effects of major power support can be translated to Mediterranean relations (ibid.: 293-4). External support alters regional balances in favour of the client, often giving the dependant a temporary or sometimes decisive edge over its rivals in the vicinity, as the Arab-Israeli situation demonstrates. External assistance encourages the client's regional rivals to seek help from alternative patrons to check against the emergence of a regional power centre, as the Cold War patron-client relationships between Syria and the Soviet Union and Israel and the United States demonstrated. A constant supply of external assistance often intensifies regional rivalries and can also provoke a regional arms race, if not outright conflict. The U.S.-Israeli partnership and Soviet support for Syria was again characteristic of this trend. External military assistance can also tempt regional leaders to attack a neighbouring country, as Libya did in the case of Chad.

Outside support prolonged stalemates throughout the Cold War as rival major external powers sought to preserve the regional status quo. The Cold War also witnessed instances when "clients" had substantial leverage over their patrons. Soviet relations with Libya and Syria and U.S. squabbles with Greece and Spain over military base rights illustrate this point.

East-West detente has forced regional leaders to reassess their sources of external support. The change from a bipolar to a multipolar international system creates a more advantageous situation for external actors in the region. Their assistance becomes more effective in influencing regional dynamics.

That is, the more competitive environment for international capital in the 1990s means that only those regional actors who are deemed as politically stable and economically productive will be extended the lines of credit required to make a difference to their overall international position. The conditions attached by the IMF to its debt-rescheduling package for Algeria in mid-1995 demonstrates this fact clearly (Financial Times, 13-7-95: 3). The countries of the Maghreb region are currently attracting less foreign capital than those of East Asia and even less than some others in Latin America. This is due to a number of reasons which include political uncertainty, administrative obstruction, a comparatively unskilled labour force, and an inadequate infrastructure (Strange, 1995: 249-50).

Arms transfers also affect domestic politics, benefiting those sectors of society obtaining large consignments of military hardware. However, too high a profile of military relationships with major external powers can sometimes give rise to an increase in domestic resistance activities as the Algeria-France and Morocco-U.S. relationships illustrate. Substantial shifts in domestic political regimes are also frequently followed by changes in alignments with major powers, as post-Cold War relations between Malta and Western Europe portray (Calleya, 1994a: 138-47).

The disappearance of the Soviet Union leaves the U.S. as the predominant external military actor in the Mediterranean, and has allowed it to consolidate its position in the area. The presence of a single power could help to moderate local crises, as Washington mutes rivalries by cutting off supplies to mavericks in the basin. As an economic hegemon in the area, the European Union could assist the U.S. in this sector by complementing American military power with economic support. But such co-ordination will not be easy to achieve given the more competitive nature of trans-Atlantic relations in the post-Cold War era.

At a bilateral level, the U.S. and France remain the two dominant great power states in the Mediterranean. The U.S. is the leading external actor in the area with strong political, economic and military ties to its Southern European allies in NATO. It also has comprehensive agreements with Israel, Egypt, Cyprus, Malta, Tunisia and Morocco. The sheer economic and defensive power that the U.S. possesses ensures that it will continue to attract the attention of Mediterranean countries at the start of the next millennium. While the post-Cold war period has seen the Americans become more concerned with events in the eastern sector of the basin, French links in the Mediterranean remain predominant in the western sector. As an external great power the United States can act more independently in the area than the internal great power of France which is much more vulnerable to retaliation from action in the Maghreb given its geographical proximity to the area and its large Maghrebi emigrant community. Nevertheless, both powers influence the balance of power within the subregions bordering the Mediterranean as U.S. support for Israel and French assistance to the Algerian authorities makes clear. Both powers also fuel perceptions of neo-colonialism across the Southern shores of the Mediterranean, a hardly surprising finding given the asymmetry between the economic and military presence of the Americans in the Levant and the French in North Africa and that of other powers, such as Britain and Germany. In sum, the key distinction between internal and external great power roles in international regions is that the latter is much more physically detached from the area in question, and in theory, can therefore adopt a more unilateral stance when it comes to taking action. The Mediterranean case tends to reinforce this observation especially if one compares the much more flexible approach the Americans have been advocating vis-à-vis the Algerian crisis than that pursued by the French, particularly during the initial stages of the civil war (Financial Times, 13-7-95: 3). In any case, the internal and external great power distinction does not change the basic fact that intrusive actors can influence regional dynamics but they cannot dictate patterns of relations within international regions. The U.S. experience in Lebanon and the more recent French endeavours in Algeria underline this point.

More limited bilateral contacts occur between Mediterranean states and individual European Union member states. Spain and Italy have comprehensive political and economic links with Morocco and Tunisia respectively. Mediterranean countries also have commercial links with Germany and Japan, and are attempting to supplement foreign direct investment from these donors with capital from the Asian power-houses of Taiwan, Singapore and South Korea. To date such efforts have had little success. The entrance of Central and Eastern Europe on to the investment circuit coupled with the increase in domestic instability across North Africa has made it more difficult to obtain funding from potential investors than was the case during the Cold War.

Trends in the aftermath of the Cold War suggest that the roles external powers will play in the Mediterranean will differ in important ways from previous phases of history. In areas of marginal interest, the disappearance of Cold War zero-sum competition for global influence is likely to lead great powers to disengage from those states disinterested in resolving internal conflicts and introducing reforms to promote democracy and the respect of human rights (Lyons, 1992: 196). The challenges confronting North African countries fall within this taxonomy. Unless states such as Algeria attempt to carry out the reforms cited above, external powers are more likely to opt out becoming involved in the crisis.

Operations such as the U.S. campaign in Somalia in 1992-93 and the French expedition in Rwanda in 1994 have shown that external powers, including the last superpower, can at best encourage humanitarian policies. Yet, it is the internal dynamics and actions of regional governments and non-government leaders that will ultimately control the timing and success of such reforms (ibid.).

To summarize, here is a synopsis of the types of great power actions and their influences on the development of international regions. External support has sometimes helped weaker states preserve their security interests and acted as a deterrence against regional hegemons. The positions of Israel and Egypt in the 1950s illustrate this trend in the eastern sector of the Mediterranean. In certain situations, external powers can help to moderate local tensions. In the eastern sector of the Mediterranean, where rivals Greece and Turkey both receive assistance from a single power - the U.S., the intensity of their rivalry has been muted by the diplomatic activity of their supplier. More recently, the European Union is also assuming a similar role in the area, as demonstrated by its intervention in the Cypriot affair.

External support often encourages a client's regional rival to seek assistance from an external power rival of its neighbour's patron: for example, Syria linked itself with the Soviet Union to counter the United States' ties with Israel. In contemporary international relations several clients are having to compete for financial assistance from similar donors: Israel's and Egypt's position vis-à-vis the U.S. perhaps best illustrates this trend.

The availability of assistance from major power rivals can intensify regional tensions. For example, large transfers of military supplies to the Middle East helps to fuel the arms race in this region and could again provoke an outbreak of hostilities. Arab-Israeli conflicts illustrate this reoccurring pattern of relations. Current ties between states in the Mediterranean and great powers such as the U.S., Russia and China can therefore lead to an escalation of regional tensions.

External support can assist in prolonging stalemates. The continuous supply of economic and military supplies removes the necessity to offer concessions to regional rivals. The Moroccan-Algerian stalemate in the Western Sahara illustrates this tendency.

External support is sought by both the weaker and the stronger states in an international region as a way of preserving and improving their relative position. This helps explain why countries in the Maghreb and Levant have recently sought to form alliances with the West (Egypt, Morocco and Tunisia with Western Europe and Jordan with the U.S.).

External assistance can also help bolster the political faction receiving supplies. Military regimes are thus able to consolidate their positions, although too high a profile with external actors could result in a domestic backlash. The cases of Spain and Greece in the 1970s is reflective of this trend. Egypt's ties with the U.S. in the 1980s and early 1990s also depicts this pattern of relations.

Occasionally, clients find that they have substantial leverage when dealing with patrons who were competing for the allegiance of smaller states to help consolidate their spheres of influence. The American experience with Israel and the Soviet experience with Libya reflects this tendency. The end of the Cold War has seen the number of potential client-states multiply, thus diminishing their overall bargaining power.

The autonomous nature of indigenous regional dynamics is reflected by the fact that changes in domestic political regimes are frequently followed by changes in alignments with external powers. Such shifts occur to serve the clients own interests. The shift in Malta's foreign policy since 1987, when the Nationalist Party took over after sixteen years of a socialist government and aligned itself with the West, is indicative of this point.

The state of flux that the international system has been in since the collapse of the Cold War, has led all the actors in the international system, including the great powers, to be much more flexible in their foreign policy endeavours than during the period 1945-89. Such flexibility is another factor regarded by many theorists as a hallmark of effective great power concerts (Kegley, Jr. and Raymond, 1994: 218-20). In situations where the direct interests of the nonregional states are concerned, they will react decisively as demonstrated by the U.S. in the Gulf War 1990-91. The collapse of the Cold War has reduced the successor states' ability to intervene in regional affairs. Indeed, Russia is having to dedicate the majority of its political and military resources to securing its own borders, as the Chechnya crisis highlights. As Moscow reduces its support to countries in and around the Mediterranean, Washington may decide that it is no longer necessary for it to maintain an extensive network of contacts in the Mediterranean. On the other hand, it can also be argued that Washington would face fewer risks of countervailing opposition if it chose to intervene more often.

In international regions where external powers' interests have largely been derivative of their mutual rivalry, one can expect a decline in their intrusive actions. The implosion of the Soviet Union has already led to U.S. military reductions in Western Europe, a trend which is likely to continue. A reduction in superpower overlay will allow regional dynamics to function even more independently. Other things being equal, it can be expected that in asymmetrical international regions, such as the Middle East, aspiring pre-eminent regional power centres will pursue hegemonic goals, and the smaller states in their vicinity will be more prepared to acquiesce. In reality, things are never equal. The collapse of the Soviet Union upsets the foreign policy establishment of many countries in the Mediterranean area which benefited from Soviet support. Without Soviet support, long-time rejectionist countries such as Libya and Syria, could adopt more collaborative regional policies. Without the luxury of any serious alternatives, regional powers are likely to seek closer economic and political relationships with the U.S. and Western Europe.

The diminution of Soviet influence in the Mediterranean area has resulted in a gradual increase in regional ties with the West in general, and the United States in particular. Countries such as Egypt, Jordan, Greece and Turkey already benefit from monetary and defence arrangements with Washington. Morocco, Algeria, Tunisia, Malta and Cyprus are also seeking closer ties with the European Union. However, continued domestic instability in many of these countries is impeding the attainment of far reaching agreements with great powers. Future external relations in the Mediterranean area will depend less on the activity of the nonregional powers and more on how riparian states consolidate their power and perceive their geostrategic interests. For example, arms flows to the eastern and southern sectors of the Mediterranean basin are likely to continue at a steady pace in spite of events like the Arab-Israeli peace process. Regional powers in these Middle Eastern subregions will be even more determined to shore up their defences now that patron support can no longer be taken for granted.

The termination of the Cold War has therefore forced regional powers to reassess their foreign policy aims. One outcome that is already noticeable is that the regional challenges of the future lie more in the economic and financial battles for markets than the military battles for territory. One incentive for regional states to end border conflicts (Arab-Israeli, Greece-Turkey-Cyprus) is so that they can participate in international financial markets without any restrictions. The globalization of economic activity, which is characterized by the growing frequency of cross-border transactions, the ever-increasing volume of international trade, the growing strength of international investment, and the enhanced complexity of the international division of labour, is therefore likely to increase centrifugal tendencies throughout the international system, and especially in less developed areas such as the Mediterranean.

An analysis of external involvement at a bilateral level shows that major powers responses to regional invitations and their initiatives on behalf of their mutual rivalry can have a significant effect on regional patterns of interaction. What they have not done, however, is to alter the basic pattern of regional alignment and conflict within these international regions. As indicated in the framework of analysis, a review of great power roles in regional relations is essential because it is their involvement which makes these units of analysis international.

7 The Role of Multilateral Organizations

7.1 Introduction

Chapter seven looks at the influence of international organizations with a power projection capability to influence regional relations in the Mediterranean. An explanation and assessment is given of the various Mediterranean policies that have been proposed by the European Union (EU), the Organisation on Security and Co-operation in Europe (OSCE) and the North Atlantic Treaty Organization (NATO). The political interests and coalitions that lie behind these international organizations, as well as the role these arrangements play in the Mediterranean area are examined.

Freedom of navigation has been a principal concern for all external actors who have an interest in the Mediterranean. Historically, the sea has been the chief medium for cultural and economic exchanges and for political and military ventures. When the basin was controlled by Mediterranean states they based their strategic considerations on the nature of their physical position and the paramount position of the Mediterranean in the international political and economic system.

The pattern of relations between the internal and external actors in the Mediterranean changed significantly in the twentieth century once the decolonization process became irreversible. By the mid-1950s a number of international organizations had already expressed anti-colonial tendencies. In 1918 the Covenant of the League of Nations stressed the right to self-determination, and in 1945 the Charter of the United Nations reiterated this principle. Both contributed to establishing an international consensus which was hostile to the possession of colonial territories. By the 1960s, Third World nationalism had gained in prominence, and the two superpowers had little choice but to accommodate a third force in international affairs, once the non-aligned conference was held at Bandung in 1955. Given the

187

intersection of the East-West and the North-South divides in the Mediterranean, it perhaps comes as little surprise that two of the founding fathers of non-alignment, Tito and Nasser, came from two countries in the Mediterranean area (Fenech, 1993: 136).

The international system underwent another significant change in structure in 1989 with the end of the Cold War. What impact this development will have on intrusive activity in regional relations remains unclear. The fear in some quarters, that the United States might become a global hegemonic power, has subsided as Washington has adopted a more selective foreign policy of limited engagement. The possibility of a "back to the future" course of events emerging, with great power patron-client relationships of the past resurfacing, has also not appeared. The emergence of a "new" hegemonic actor on the international scene also remains an illusion. One certainty is that international organizations are playing a more active role in regional politics. The relegation of superpower politics to the history books has coincided with an increase in multilateral intergovernmental and transnational patterns of interaction.

The end of the Cold War has seen a significant change taking place in the realm of external actors' ability to influence international regional relations. Bilateral types of intrusive intervention are being replaced by multilateral types of involvement, as international organizations become more active in regional patterns of interaction. This is evident if one compares the nature of intrusive action during the Cold War in the Mediterranean area, with that of the 1990s. The disappearance of the Soviet Union and the reluctance of the United States to act unilaterally has allowed international organizations such as the European Union and NATO to play a much more active part in the Mediterranean space. A number of other indicators also appear to support this thesis. First, the multifaceted security challenges that great powers perceive as emanating from this area are convincing them that international organizations are better equipped to contend with such risks. Second, the costs of confronting such challenges favours a collective intrusive response which shares the economic burden of such action. Third, the Mediterranean remains a geostrategic area of importance, both as an international waterway and because of its petrol producing capacity. It is therefore in the interest of all international actors that the sea-lines of communication in the Mediterranean remain open. A multilateral approach to such security challenges is less of a political risk than unilateral action would be.

Although the United States remains the predominant military actor in the Mediterranean basin, U.S. cutbacks coupled with the reactivation of WEU military schemes, particularly between France, Spain and Italy, signals a period of a more evenly distributed, if still very uneven, military balance among the intrusive actors. This development is also conducive to enhancing the longevity of international organization involvement in the area.

7.2 The Role of International Organizations

Several international organizations operate in the Mediterranean area. Seen from a post-Cold War perspective, these interfacing and occasionally overlapping multilateral institutions can be arrayed on a continuum of varying size, composition, and purpose (see Figure 7.1).

Large Membership   Limited membership
Multidimensional   Limited objectives

UN - OSCE - Arab League - AMU - EU - WEU - NATO

Figure 7.1 - Spectrum of International Organizations in the Mediterranean Area

At one end of the continuum is the United Nations, a multidimensional intergovernmental organization with almost universal membership. In addition to issuing resolutions concerning crisis situations in the area, recent years have seen a multiplication of UN peace-keeping missions around the Mediterranean basin (New York Times, 6-1-95: 10). Next on the continuum is the Organisation on Security and Co-operation in Europe. A smaller organization, the OSCE possesses a much more limited mandate than the UN. It is mainly a talking shop where security issues can be openly discussed. Although some Mediterranean countries are members of this institution, the majority are not which qualifies it as an intrusive actor. The Arab League, which includes the Arab Maghreb Union (AMU), is the next multilateral institution on this continuum. Its large size has not been matched by effective measures which have influenced the course of events in the Mediterranean. For this reason and the fact that both the Arab League and especially the AMU appear to be in a state of paralysis, they are not included in the analysis that follows. The European Union and the WEU are next along the continuum. Its membership of Mediterranean countries is far larger than that of the OSCE and it is therefore better described as a quasi-intrusive actor. All the countries in the Mediterranean basin are economically dependent on the EU to the extent that the EU is already an economic hegemon in the area. Since the launching of its "Global Mediterranean Policy" in 1972, the EU has implicitly treated the Mediterranean as its sphere of influence in both an economic and political sense. The admittance of Greece, Spain, and Portugal during the 1980s strengthened the EU's links with the Mediterranean. Any progress towards deepening the process of European integration would equip the EU with a more multidimensional approach in the Mediterranean. Finally at the opposite end of the continuum is NATO. It comprises fewer members from the Mediterranean and has a limited political and military range of activities. In some instances these multilateral mechanisms overlap, though they are not necessarily incompatible (Kegley, Jr. and Raymond, 1994: 220).

In summary, this chapter provides a better understanding of the influence international organizations have on regional dynamics operating in the Mediterranean. It also aims to clarify what relationship exists between international organizations and international regions. For example, what interests are international institutions seeking to promote by intervening in regional relations: a regional interest? a great power interest? or some independent self-referenced interest? Taken together, chapters six and seven attempt to provide a clear picture of the impact both bilateral and multilateral types of intrusive action can have on regional development. This will assist further in identifying what the prospects for regional development are in the Mediterranean in the post-Cold War era.

7.2.1 The Role of the European Union

7.2.1.1 The Global Mediterranean Policy

In 1972 the European Community [(EC) which became the European Union (EU) in 1994] launched its first scheme in the Mediterranean, the so-called Global Policy. Its main aim was to establish a series of preferential trade and co-operation agreements with non-member Mediterranean states along the northern sector of the basin. Association agreements had been signed with Greece in 1962, Turkey in 1964, and Malta in 1971. By 1973 agreements were also signed with Cyprus, Yugoslavia, Portugal, and Spain. The Commission also commenced a series of agreements with the Arab Mediterranean countries and Israel.

These association agreements were intended to be the first step in a process leading to a customs union and eventually full EC membership. All of the accords established free access to EC markets for most industrial products, albeit on different time scales. Access to agricultural commodities was facilitated, although some tariffs remained. The EC imposed quotas on refined petroleum, cotton, and phosphate fertilizers for a transitional period. The principle of reciprocity (the granting of preferences in return) was not applied immediately in all co-operation agreements. States experiencing a period of economic decline were even entitled to take protective action until their economy recovered. Most of the agreements with the EC were accompanied by financial protocols indicating the amount of assistance the Mediterranean country would receive in each category. Financial aid took the form of direct grants, as well as loans from the European Investment Bank (EIB).

European Community communiqués concerning the Mediterranean, consistently underline the fact that Europe's long-term objective is to form a comprehensive partnership with its neighbours in the South. In actual fact, the EC's main aim is to consolidate its sphere of influence in an effort to keep a check on potential flashpoints in the area. The EC's political rhetoric of partnership has been overtaken by the political reality of inaction. The EC's attraction to the basin stems from the member countries' interests in six major problem areas:

i)   the commercial aspect, particularly in regard to export markets for the Community;
ii) the security of energy supplies for Western Europe;
iii) the security aspect, especially the southern flank of NATO operations and possible spill-overs from the Arab-Israeli and other Middle Eastern crises;
iv) political stability in the Mediterranean countries which equates to a policy of preserving the status quo;
v)   sustaining traditional cultural-historical links;
vi) maintaining a process of dialogue in the Mediterranean area to contend with whatever challenges emerge (Regelsberger and Wessels, 1994: 246).

In 1982 the European Commissioner with special responsibilities for Mediterranean affairs, Vice-President Lorenzo Natali, delivered a speech about the need for a coherent and positive approach to the Mediterranean:

Geopolitical reasons in themselves make an impressive case for the necessity of a coherent European Community policy on the Mediterranean. A glance at the map proves it. Look first at the Balkans and the mouth of the Atlantic. Take in the Dardanelles and the petrol producing region of the Near East: remember too that the Mediterranean is the inescapable north-south axis for links between Europe and Africa. We must question whether the Community could survive a serious disturbance in the Mediterranean...(Blacksell, 1984: 285).

A review carried out by the European Commission in 1982 revealed that the EC Mediterranean Policy was far from achieving its main goal of establishing a free trade zone with its southern neighbours (Commission of the European Communities, 1984; see also Tsoukalis, 1977; Pomfret, 1986). Agricultural produce which the Mediterranean countries sought most to export, (e.g. citrus fruits, olive oil and wine), were already in surplus in the Community. In addition, instead of providing a market for industrial goods from the South, the Community had been obliged to protect its own manufacturers against competition, especially in the textiles, footwear and processed foodstuffs sector. The 1982 review led the European Commission to draw up an integrated plan for the development of its own Mediterranean regions and to adopt a new policy towards the non-member countries of the basin. One of the policy's principal aims was to help Mediterranean states overcome their dependence on imported food by helping diversify their agricultural production. Recent statistics show that this initiative has had a somewhat positive impact on certain countries but not on the Mediterranean area as a whole. For example, 1991-92 figures portray Tunisia as enjoying a positive food balance for the first time in two decades (Financial Times Survey, 14-6-93: V).

The review also reiterated the principle of free access to the Community market for industrial goods originating in the Mediterranean and an increase in financial assistance to encourage the complementary development of the different economies of the partner countries. In 1985 provisions were also introduced to ensure that non-member Mediterranean states would not be adversely affected by the accession of Portugal and Spain to the Community in 1986. In reality, the inclusion of these Southern European countries into the EC deepened further the marine frontier between the Western European and Middle Eastern international regions. By the latter part of the eighties, two factors in particular undermined the credibility of the EC "global" Mediterranean policy. One was the fact that Southern European enlargements made it even more difficult to penetrate EC markets than before. The heavily agricultural and semi-industrialized economies of the new southern members acquired first claim on the European market at the expense of non-members with similar economies (Pomfret, 1986: 98-100; 1989: 11-7). The other factor was the Single European Act of 1986. The implementation of this phase of European integration saw a proliferation in the amount of non-tariff barriers, quota systems and protectionist measures in trade dealings between the EU and the rest of the world. All these developments reinforced the notion of an emerging "Fortress Europe" (Fenech, 1993: 137).

In December 1990 the EC decided to introduce its "New Mediterranean Policy" in an effort to support the gradual movement towards economic liberalization and democratization. The new policy comprises six main components:

* assistance in the process of economic adjustment;
* encouragement of foreign direct investment (FDI);
* an increase in bilateral and Community financial assistance;
* strengthening arrangements governing access to the Community market;
* inclusion of the Mediterranean area in the Community's single market programme;
* increasing economic and political dialogue at a regional level whenever possible.

The 1990 Mediterranean policy provided for an overall package of ECU 4,405 million, sub-divided as follows:

* ECU 2,075 million in loans and grants for the countries in the Maghreb and Levant over a five year period from November 1991. This amount includes support for structural adjustment programmes undertaken in conjunction with the International Monetary Fund (IMF) and the World Bank;
* ECU 2,030 million for more broadly based financial co-operation, with a particular emphasis on promoting investment, developing small and medium sized businesses, and protecting the environment;
* ECU 300 million as a back-up for economic reform. This fund was set up to help compensate for the adverse social effects of adjustment programmes (for instance, the effects on those below the poverty line of a reduction in subsidies for essential commodities). In July 1991 an additional ECU 60 million was granted to the Palestinians in the Israeli-occupied territories who had been adversely affected by the Gulf War in the preceding

months (Joffe, 1993: 244-5; Commission of the European Communities, 1991: 4-8; see also Spencer, 1993: 52; Allen, 1978: 323-42).

Nevertheless, recent data discloses the EU's failure to embrace the necessary long-term policies required to create an all encompassing forum for tackling the increasing social and economic disparities that exist between the two international regions which encompass the Mediterranean (Pace, 1993: 7; Roseta, 1993: 17-8). For example, a comparative study of the levels of development on the two sides of the Mediterranean reveals a ratio of one to ten, which is still widening. This fact was recently reiterated by the European Parliament which deplored the fact that, in economic terms, the gulf between the two shores of the Mediterranean is on the whole growing and that twenty-three years of co-operation have not succeeded in reducing this trend (Cerretti, 1993: 4).

While economic growth is being experienced by most of the non-member Mediterranean states, including the countries in the Maghreb, growth is insufficient to provide for an improvement in living standards throughout the area due to the constantly increasing population figures. It seems that the only way to prevent the resurgence of instability in the Mediterranean is by holding out the prospect of anchoring the non-member Mediterranean states and the Levant and Maghreb subregions to the Western European international region in the long-term. At least three approaches, currently underway, aim at achieving this objective. Firstly, there is the preliminary accession negotiations with the three Mediterranean applicants, Cyprus, Malta and Turkey. Second, is the Euro-Maghreb Partnership policy reformulated at the end of 1994 in the Euro-Mediterranean Partnership document and in November 1995 in the Euro-Mediterranean Barcelona document. Thirdly, are the still evolving policies towards adjacent hinterlands in turmoil, namely the Balkans and the Levant (Calleya, 1994c: 167-88).

Relations between the EU and the non-European Mediterranean states acquired a new twist with the end of the Cold War. The termination of the East-West contest removed the political reason for admitting new states from the South into the EU. Moreover, the Community's speedy handling of the Austrian, Swedish, Finnish and Norwegian applications and its swift implementation of European Agreements and the PHARE programme with Eastern and Central European states has fuelled the chorus of discontent among EU aspirants in the South. Brussels's claim that the EU's concern is to strike a fair balance between the north, east, and south, and at the same time highlight the Union's presence in the Mediterranean has mostly fallen on deaf ears.

7.2.1.2 European Union Relations with Cyprus, Malta and Turkey

In the case of Cyprus, the EU has done its utmost not to become directly involved in the Cypriot stalemate. It has instead opted to support whatever measures the UN has adopted to resolve the issue. The EU has also continued to adopt a consistent line in its bilateral negotiations with Cyprus in the framework of the association agreement signed in 1972. Two principles have been constant throughout: that there is only one legitimate government of the Republic of Cyprus and that is the Greek Cypriot government; notwithstanding the first principle that the benefits of EU association should accrue to both communities on the Island (Redmond, 1993: 18; see Redmond, 1993).

In its Opinion Report on Cyprus's application to join the EU at the end of June 1993, the Commission delivered a positive message to the divided Island as an incentive to try and break the Cypriot deadlock. It emphasized that accession negotiations could commence as soon as there was sufficient certainty about the prospects for unification of the Island. The Commission then added the proviso that if no prospects for agreement were in view by the beginning of 1995, it would review the situation and assign responsibility for the failure. In February 1995, the EU shifted its position vis-à-vis Cyprus. In return for not vetoing a customs union with Turkey, the EU promised Greece that it would commence EU admission negotiations with Cyprus within six months of the end of the 1996 intergovernmental conference (Financial Times, 2-2-95: 2). The nature of the tactical diplomacy employed by the EU in this instance demonstrates the growing political and economic power this international organization espouses in the area. It also reflects the EU's potential to navigate regional dynamics in the Mediterranean towards co-operative settlements.

Malta submitted its EU application in July 1990, and has since launched a series of legislative and economic measures as part of a wider process of developing and adopting the Community's acquis communaitaire. In its Opinion Report on Malta's application to join the EU in June 1993 the EU highlighted the economic anomalies between the EU and Malta as a major stumbling block that would have to be resolved before accession negotiations could commence. After drawing up an "adjustment protocol" supported by technical assistance from the Community, the Maltese Government has dedicated its political and economic resources to introducing and implementing legislative reforms cited in the AVIS. In February 1995 the EU announced that accession negotiations with Malta would commence within six months of the end of the 1996 intergovernmental conference (IGC).

During the IGC the EU has to resolve the institutional implications of having a Union with up to twenty members. Second, it has to decide what rules will govern those small states that have been labelled micro-states. Third, the EU has to clarify its position on the question of neutrality and non-alignment. It remains to be seen whether such contentious issues can be resolved at the 1996 gathering.

Turning this analysis around, what are the principal reasons why Malta is seeking full EU membership? The first reason is economic. Three quarters of Malta's trade is conducted with the EU. Although the Maltese Island benefits from preferential trade agreements, full membership would guarantee access to the European market, even in times of recession. Second, inclusion into the EU would make Malta a more attractive location for foreign direct investment. After a sporadic phase of nationalization in the 1970s under the leadership of Dom Mintoff, international investors would welcome an EU anchor which advocates free trade. Finally, perhaps the most important reason spurring Malta to attain EU membership is that of security. Sixty miles south of Sicily and two hundred miles off the coast of North Africa, Malta finds itself on the faultline of two completely different international regions. Rather than risk being caught in a security vacuum, Malta is seeking to tie itself to the Western European international region. If admitted into the EU, Malta would become the EU's furthest border south, a critical location if interaction between the two international regions in the area were to suddenly become conflictual.

Turkey formally applied for full EC membership in April 1987. Immediately two basic problems were identified as major obstacles inhibiting any rapprochement between the two sides: the Cyprus issue in the Council of Ministers and the human rights issue in the European Parliament.

The Commission's deliberations over Turkey's application to join the EC centred upon four main areas: economic, political, strategic and cultural. In spite of recent economic growth, Turkey is identified as essentially a relatively poor underdeveloped country. As a result, the financial cost of assisting Turkish accession is regarded as substantial and would impinge upon the budgets of all the Union's main sources of revenue: the Common Agricultural Policy reserve, the European Investment Bank, and the Cohesion Fund. The principal political issues hampering closer EU-Turkish relations remain Turkey's position vis-à-vis Cyprus, and its ambiguous attitude towards

Greece. Other concerns include the nature and development of Turkish democracy and its respect of human rights, particularly their handling of the Kurds. Opponents of Turkish EU membership also cite the cultural differences between the Western Europe and Turkey. With religion as their basis, they reiterate the historical enmity between Europe and the Ottoman Empire and the more recent fear of a resurgence of Islamic fundamentalism. Although such fears seem excessive, repetition of such arguments does nothing to narrow the differences of policy-makers in both Ankara and Brussels.

The one area which certainly advances Turkey's case for EU membership is the strategic one. Its geographical location and its membership of NATO makes Turkey a buffer between a stable Western Europe and an erratic Eastern Europe and Middle East. Turkey's role in the Gulf War re-established its position as an essential component in the Atlantic Alliance at precisely the time when the end of the Cold War augured for it to be relegated to the side-lines.

The Commission issued its Opinion Report on Turkey's application on 20 December 1989, summarizing the negative themes outlined above ("Commission Opinion on Turkey's Request for Accession to the Community", Commission of the European Communities, 1989). As an alternative to full membership it proposed a revised and more comprehensive association agreement. In June 1990, the Commission announced full details of what it had in mind: the completion of the EU-Turkey customs union in 1995, the promotion of industrial and technological co-operation, the resumption and extension of the financial protocol and the reactivation of political and cultural exchanges. Turkey has subsequently recognized that its goal of EU membership will have to become a long-term objective. In the interim, Ankara has concentrated its diplomatic resources on establishing a customs union with the EU, which came into effect at the end of 1995. Greece's support on this issue was won over in return for an EU announcement that it would commence membership negotiations with Cyprus within six months of the end of the 1996 IGC, i.e., the same condition that has been offered to Malta. The ability of the EU to consistently have a positive effect on Greek-Turkish relations is however not something that can be taken for granted, as the Aegean Islands stand-off in February 1996 demonstrated. The EU's expression of solidarity with Greece somewhat alienated Turkey and served notice that in times of crisis the EU would take the side of its member states (International Herald Tribune, 13-2-96: 8).

7.2.1.3 The Euro-Maghreb Partnership Policy

The European Union's second type of outreach programme towards the Mediterranean complements the first but stops short of actually holding out the prospect of accession. The communication from the Commission to the Council and the European Parliament entitled: The Future Of Relations Between The Community And The Maghreb, announced the new concept of establishing a Euro-Maghreb partnership. This model was to replace the development of a co-operation policy which the EU had previously been offering (Commission of the European Communities, April 1992: 5).

The EU's aim in launching this new concept is multidimensional. The realities of geographical propinquity, economic interdependence and population movements are just some of the indicators that have convinced Brussels that it is in its best interest to strengthen links with countries in North Africa. The EU's principal aim is to deliver a reassuring political and economic message to the Maghreb. As if to emphasize that it realized that its past initiatives were quickly being superseded by international developments, the EU stipulated that its ultimate goal is to establish a Euro-Maghreb economic area (ibid.; Time International, 14-6-93: 36-7).

A number of key elements essential to the economic restructuring of the Maghreb were identified and integrated into the Union's new strategy towards this Middle Eastern subregion. These include schemes to foster economic reforms, increase investment, develop trade liberalization, contain population growth, and enhance political pluralism. The execution of these objectives would therefore increase Euro-Maghreb intergovernmental interaction at both a political and an economic level.

Direct foreign investment statistics for the period 1982-92 indicate the urgent need to overhaul the system to create an environment much more conducive to economic activity. Although Algeria, Morocco and Tunisia are all at different stages in their structural adjustment programmes, the Union formulated a series of support schemes to assist each of these countries through this process. These include technical assistance in sectors such as fiscal and financial reforms and restructuring businesses. Support is also to be extended to sensitive social sectors such as health care, education and housing programmes, where the most strict budgetary measures would be too momentous to handle without safety nets. Direct support for schemes linking vocational training and job creation such as privatization and the creation of small business ventures was also included. These measures would therefore assist in strengthening the weak Euro-Maghreb intergovernmental links that exist at a socioeconomic level and could help to promote transnational patterns of interaction across the Mediterranean.

An essential feature of the partnership concept envisaged by the Union is the complementary development of both vertical and horizontal integration, which to date has been substantially lacking (Strange, 1995: 250). Although the first attempt at setting up a single Maghreb institution dates back to 1958, most of the subregion's diplomatic efforts and resources have been consumed in coping with the implications of independence and regional disputes (see section 4.3.3 for assessment of intra-Maghreb relations).The EU adopted a series of measures in the partnership document to encourage the process of horizontal integration, i.e. Maghreb integration, is a central point which assists in clarifying the nature of regional dynamics operating in the area. The fact that the Western European international region is promoting further integration in the subregion (the Maghreb) of another international region (Middle East) indicates that EU assistance is not tantamount to EU acceptance. In simpler terms, an increase in EU economic assistance to the Maghreb does not signify that the former necessarily wants to further the process of rapprochement between the two international regions bordering the Mediterranean. On the contrary, the EU's main aim appears to be to maintain the distance, if not the disparity, between the two sides. After all, EU membership is not on the cards, as Morocco discovered after submitting its application to join the EU in 1987.

With very limited success to talk about in its economic endeavours in the Maghreb, apart from investment ventures in the energy sector, the EU has had very little success in its political and social programmes in this Middle Eastern subregion. The Union claims that it aims to enhance the trend towards political liberalization in the Maghreb by linking economic assistance to democratic reforms. Yet no such linkage has taken place. Human rights abuses and a general lack of liberal principles in society remain. The extreme case of Algeria highlights this point clearly.

The EU has also not developed a comprehensive policy which addresses population growth and its potential impact on migratory trends. While estimates vary, most forecasts predict that the total population of the Maghreb will double by the year 2025 (Joffe, 1994: 33-6). Even if the Maghreb is included in a free trade scheme mirroring the European Economic Area (EEA), the need for a pan-European immigration policy will still be required if the potential threat of mass migration, triggered by declining living standards and rising unemployment, is to be contained.

The EU's policies in the immigration sector underlines the thesis that the notion of a Mediterranean international region is either fictitious or irrelevant. Since the introduction of the Single European Act, EU immigration policy has become even more restrictive, invoking its description as a "fortress Europe". Intergovernmental social patterns of interaction are therefore very limited. There is no intention of imitating schemes like the North American style points system of qualification for immigrants and no attempt to identify Europe's probable economic needs and capacity to absorb immigrant labour in future decades (Mortimer, 1992: 39). Progress in these areas would signal that the EU is serious about making the Mediterranean a drawbridge between Western Europe and the Middle East, rather than a moat. Yet, the fostering of a Euro-Middle East international region will require much more than dealing with the symptoms caused by Euro-Middle East disparities. The West European and Middle East international regions will not move closer together until the former is prepared to consider allowing the latter to participate as equals in the European international region. Without the opportunity to participate at all levels of competition, countries in the Levant and the Maghreb will lack the incentive and belief in the EU's often cited long-term objective of anchoring the southern and eastern shores of the Mediterranean in a technical, economic, political and social sense.

The limited nature of EU policies in the Mediterranean and the absence of an overall strategy is evident when one assesses European energy ventures in the Maghreb. As energy is one of the few privileged sectors for horizontal co-operation in the Maghreb subregion, a large proportion of the ECU 1.8 billion available from the EIB is directed towards investment in this field (see Mikdashi, 1995: 37-64, Luciani, 1995: 65-87, and Ahmed, 1995: 111-52). In addition to the pipeline linking Algeria and Italy via Tunisia which has seen its throughput capacity double since it came into service in 1983, preparatory work on a second gas pipeline from Algeria across Morocco and the Straits of Gibraltar to Spain and eventually France is already underway (Ghiles, 1992: 17-8).

7.2.1.4 European Union Policies in the Levant and the Balkans

The European Union's third type of outreach programme which is directed towards the Levant and the Balkans is largely conjectural. It is dependent on isolated political, military or economic developments. The reasons for a lack of progress in policy formulation in this sector are numerous. The role of the United States and the importance attached to differences in policy priority are two factors contributing to the impasse in policy-making in this area. EU preoccupation with traditional security issues does not fit in with the priority non-member Mediterranean states attach to economic and development issues. All of these factors have prevented the emergence of a constructive dialogue between the EU and its Mediterranean neighbours.

The EU's indecision and reactive approach during the initial stages of the Balkan conflict has revealed what little experience the EU actually has when it comes to conflict prevention and peace-keeping measures. Until the second pillar of the Maastricht Treaty is implemented, i.e., the common foreign and security policy mechanism, the EU remains ineffective in executing common military action policies. In short, the Gulf War of 1990-91 and the recent Balkan conflict have drawn attention to the complexities that have to be resolved before the EU can achieve its Maastricht goal of a comprehensive and united foreign and defence policy.

In the short term, the best prospects for increased EU involvement in both these volatile locations remains in the diplomatic arena. The resumption of the Union's political and economic relations in the Levant after the Gulf War, and the acceptance of the EU's participation in the Middle East Peace Conference by all parties concerned, including Israel, is recognition of the EU's foreign policy potential. The involvement of the EU in the Dayton Peace Settlement in the Balkans also demonstrates the positive, though limited, contribution, the EU can make in these 'out-of-area' crises.

One of the fundamental obstacles prohibiting strong EU transnational or intergovernmental links across the Mediterranean is the fact that countries in the Middle Eastern international region are at very different stages of both economic and political development. In fact, the apparent incoherence of EU policy in this region may be partly explained by the incoherence and diversity of this region. In addition, the Mediterranean basin consists of poles of power on its northern, southern and eastern shorelines, which distracts riparian states from establishing stronger trans-Mediterranean links. The extent of this asymmetry between the Western European and Middle East international regions complicates further the task of Mediterranean centrists who want to foster a coherent and comprehensive harmonization process to nurture intense co-operative relations across the Mediterranean. It also makes the re-emergence of a Mediterranean international region an unlikely prospect.

In October 1994 the EU outlined its new plan to overcome the problems that have not allowed the Mediterranean to be treated as a single unit for policy formulation. The document, 'Strengthening the Mediterranean Policy of the European Union: Establishing a Euro-Mediterranean Partnership', proposes the establishment of a wide-ranging free trade area and the convening of a Euro-Mediterranean Conference in 1995 (Commission of the European Union: COM (94) 427). The establishment of free trade is to be supplemented by closer co-operation in the political and economic spheres. At first sight the programme therefore holds out the prospect of promoting co-operative patterns of relations across the Mediterranean (Barcelona Declaration, Final Version, 27/28-11-95).

Given the incoherent and diverse nature of the Mediterranean area, the EU's selective à la carte approach to the area is perhaps the most realistic one to adopt. The free trade experiment is already at an advanced stage with Israel, which has negotiated access to both EU and EFTA markets for its products (Financial Times, 17-12-94: 4). Negotiations on free trade between the Union and Morocco and Tunisia have also already begun, with no parallel progress in any of the other Maghreb states. Turkey has just negotiated a customs union with the EU, while Malta and Cyprus are engaged in a process of accession to the Union. EU enlargement southwards to a select number of countries will create as many new strategic political problems as advantages and underlines the lack of a common EU Mediterranean policy. The Euro-Med document even stipulates that relations with the countries of the former Yugoslavia are beyond the scope of the communication. The immediate problem in this area is that the cost to implement the necessary peace-building measures are more than the EU on its own can afford. For example, the extra money required to finance this new policy (5.5 billion ECUs or $7 billion in EU aid from 1995 to 1999) can only be taken from existing programmes for Eastern Europe or by raising the EU's budget limits. Steps in either of these directions would certainly be opposed by the EU's main contributors, Germany and Britain. Another difficulty is that there is little consensus on the best way to help Mediterranean countries develop their economies. Several economists believe increasing aid to countries like Algeria or Egypt risks enriching a small class of profiteers or pouring capital into a bottomless pit. Proponents of the free trade blueprint argue that the answer is to open EU markets to exports from the eastern and southern Middle Eastern subregions. Such a measure would however increase competition with Europe's own agricultural produce, a development Mediterranean member states such as Spain and Italy are certain to oppose.

Given the diversity of the area, Mediterranean countries regard any EU efforts to lump them together as an unrealistic exercise. More advanced countries such as Israel and Morocco, have relatively open market economies. In contrast, countries such as Syria and Algeria still have largely state-controlled economies designed along Soviet lines, despite recent timid reforms. Countries such as Turkey, Cyprus and Malta regard themselves as on a completely different playing field from their Mediterranean neighbours, with full EU membership a realistically attainable goal. Moreover, political differences between countries in the littoral do not augur well for a period of regional rapprochement. Most of the southern Mediterranean states are not full democracies and they restrict human rights. The chaotic political situation in Algeria and the stalemate with Libya, currently in diplomatic quarantine because of the Lockerbie affair, are two cases in point that have prevented even the more modest Euro-Maghreb forum from getting off the ground.

All of these factors, coupled with the fact that the Euro-Med policy document makes no reference to creating an effective and all-embracing Mediterranean forum or organization, such as the Spanish-Italian CSCM 1990 proposal, reveals the reality of an area frustrated by the lack of a common identity. During the past two decades the EU-Mediterranean policy has been little more than an exercise in boundary management. The EU's latest proposal to establish a free trade area across the Mediterranean is an attempt to spur transnational patterns of relations to complement existing intergovernmental links. The EU's diplomatic objective is to overcome the perception shared by most riparian states of being marginalized by international organizations operating in the area. Steps towards preventing the emergence of a conflict based Mediterranean region can only take place if the central pole of power in the vicinity, the EU, acts to halt the perception that its only interest in the area is to shore up its own security interests. Efforts to increase co-operative patterns of relations between the Western European and Middle Eastern international regions must be perceived as non-discriminatory and long-term by all those involved, if the right conditions for fostering a Euro-Middle Eastern co-operative space are to be accomplished. To date, there is no sign of such a process in the Mediterranean.

7.2.1.5 The European Union's Mediterranean Agenda: An International Organization's Attempt to Navigate Regional Dynamics

As stated earlier the EU is already an economic hegemon in the Mediterranean area. All the countries in the basin are highly dependent on conducting trade with Western Europe (see Table 7.1 which indicates that only Egypt has reduced its percentage of total exports to the EU). The harmonization of economic and financial regulations achieved through the Single Market programme, and provided for in the Chapter dedicated to Economic Union in the Maastricht Treaty, augur well for a more cohesive economic hegemon that will dominate its southern periphery. Whether this process will enable the EU to play as forceful a role in the political and military spheres of Mediterranean relations, depends on how successfully it can implement its goal of establishing a common foreign and security policy as envisaged in the Maastricht Treaty. To date EU member states have shown a limited ability to pool their diplomatic and military assets into a single decision-making process. National interests continue to supersede the notion of a collective security arrangement. As a result bilateral relations, for example between France and Algeria, Italy and Libya, and Spain and Morocco, continue to dominate intergovernmental political relations.

COUNTRY	EXPORTS TO WORLD (MILL OF $)		EXPORTS TO THE EU AS A % OF TOTAL EXPORTS	
	1981	1991	1981	1991
ALGERIA	13296	12314	51.0	67.7
CYPRUS	562.4	975.2	30.2	44.1
EGYPT	3232.6	3838.2	44.2	28.3
ISRAEL	5673.3	11598.3	35.8	34.4
JORDAN	540.8	879.2	1.6	3.1
LEBANON	1001.7	490.2	4.5	22.7
LIBYA	15575	10775	52.7	83.8
MALTA	449.6	1140.7	70.4	71.8
MOROCCO	2286.5	5148.8	57.9	67.8
SYRIA	2101.9	3699.8	65.5	45.7
TUNISIA	2463.7	3826.7	61.4	73.8
TURKEY	4695.6	13334.9	33.4	50.8
FORMER YUGOSLAVIA	10929	16235	23.3	56.0

Table 7.1 - Destination of Exports to the World and the EU, 1981 and 1991
Source: IMF, *Direction of Trade Statistics Yearbook*, 1988, 1992

Trade patterns between non-member Mediterranean countries and EU member states also seem to favour the continuation of a bilateral approach in EU-Med relations (Khader, 1995: 21-35). In the 1980s, exports from northern EU member states to the southern Mediterranean area were relatively low with 2.5% from Denmark, 4.0% from the Netherlands, 4.2% from the United Kingdom, and 4.7% from Germany. In contrast, the Mediterranean presents the southern members of the EU with a dilemma: although a profitable export area, it is also a key competitor in certain markets (agriculture, textiles) in which no growth can be expected in the near future. Thus any EU concessions granted to non-member Mediterranean countries in these areas results in southern members of the Union presenting their compensation bill to the richer northern member countries (Regelsberger and Wessels, 1984: 251). Recent increases in budgetary allocations to Eastern Europe leads one to question how long the northern members of the EU would be prepared to "foot such a bill". Bilateral relations also constitute the main form of diplomacy between EU and non-member Mediterranean states. The geopolitical changes that have taken place since the end of the Cold War, particularly the emergence of Eastern Europe, has ensured that progress towards an EU-Mediterranean "co-prosperity sphere" remains fictitious.

The likelihood of an increase in EU interest in the south must also be weighed up against the recent north-east shift in EU membership. The entrance of Finland, Austria, and Sweden increases pressures on the EU to concentrate more of its resources on central and eastern Europe which is more of a direct security concern to them than the Mediterranean area. It will therefore be up to EU member states with a direct interest in the Mediterranean, such as France, Spain, Portugal, Italy and Greece, to defend Mediterranean interests at the 1996 IGC. This will be no easy task given that the EU's two largest contributors, Germany and Britain, are both adamant that economic resources should be allocated to prepare central and eastern European states for eventual membership. The EU therefore faces the difficult task of being all things to all people. If non-member Mediterranean countries are not to perceive themselves as being marginalized, then Brussels will need to introduce and implement a series of policies that are conducive to narrowing the growing disparities between Western Europe and Middle Eastern states bordering the Mediterranean (for superpower-EC positions in the Mediterranean see Lambert, 1971: 38-9).

To date, no concrete steps have been taken to commence a process of harmonization between the northern, eastern and southern shores of the Mediterranean. Even the most recent Brussels' Mediterranean programme which proposes creating a free trade area in the area by the year 2010, does not include any mechanisms to correct the wide economic gap that exists between the states in and around the Mediterranean basin. Although free trade is likely to increase the level of trade between the northern and southern countries of the Mediterranean, there is nothing to guarantee that this will reduce the level of economic disparities which currently exist. In fact, an increase in EU exports to the South would only exacerbate the negative balance of payments in this area. As a result, the creation of a Euro-Med free trade zone could end up reinforcing contemporary North-South, North-North, and South-South divides.

The EU has demonstrated that it is sticking to its "principal nations approach" in relations with the Mediterranean. Despite rhetoric supporting a global approach doctrine in its Mediterranean communiqués, political and economic relations between the EU and the Mediterranean countries are so diverse, that individual or small grouping treatment remains imperative (Regelsberger and Wessels, 1984: 259-60).

The EU's policies towards Malta, Cyprus and Turkey, the Maghreb, the Levant and the Balkans illustrate the hierarchical or preferential system Brussels adopts in the Mediterranean. The more the EU becomes obliged to set priorities in the Mediterranean in view of its other challenges - reforming its institutional mechanisms to accommodate current and future enlargements, identifying the next steps forward in achieving a single European currency and a common foreign and security policy, anchoring Central and Eastern Europe to the West, reassessing relations with the U.S.A. and Japan - the more likely the EU will be tempted to establish very strong relations with only "key-states" in the littoral. This "principal nations option" could therefore result in the formulation of a patron-client relationship in which EU predominance guarantees that Mediterranean countries remain subordinate to it as an intrusive actor.

Currently, the pattern of socioeconomic, political, and military relations between the EU and the Mediterranean are all EU dominant. Brussels continues to dictate the pace and scope of European and Mediterranean interaction as highlighted in the policy communications it has churned out since the early 1970s. In the last decade the EU has increased the degree of political and economic relations it has with its southern periphery, but has made very little or no progress at all in instigating social and military patterns of relations. The process of communication remains a one-way process, with the EU dictating the parameters of the deliberations that can take place. The hierarchical nature of the interaction indicates the central aim of the EU organization in this area: not to assist in the creation of a Mediterranean international region, but to consolidate a European sphere of influence (see Figure 7.1). In such circumstances, the southern half of the Mediterranean could become some kind of outer zone of suzerainty of the EU (see Waever, 1995: 8-15). The Mediterranean area would therefore not be a co-operative or conflict region nor a frontier zone of indifference between two distinct international regions, but an extension of the European international region. Such an outcome requires a much more concerted EU policy in the Mediterranean than has to date been the case. Despite launching the free trade plan for the Mediterranean at the end of 1994, the EU still lacks a coherent vision of long term goals in the area to be certain that the Mediterranean will revert from a superpower pond during the latter part of the twentieth century to a European lake in the twenty-first century. The sea-change in international relations since 1989 has reduced the political significance of the Mediterranean in international affairs. The end of the Cold War has removed many of the political reasons for including Mediterranean states into the EU. It also suddenly opened up a new semi-periphery to the East in direct competition with the Mediterranean. The rapid rate at which the EFTA and Austrian EU applications were processed and the proliferation of co-development policies offered to Central and Eastern Europe are indicative of this shift in EU priorities.

The probability that the EU will reverse this trend and develop an integrated policy towards all of its neighbours remains remote. Certain factors are likely to keep the countries of Central Europe high on the EU priority list. These factors include: the existence of common land borders which makes the mass migration threat from the East more extreme than that from the South, where the Mediterranean Sea acts as a formidable barrier; the significant military arsenals still stockpiled in the East which are considered more of a potential threat than the weaker military capabilities of countries along the southern shores of the Mediterranean; the opportunity to reap quicker and higher economic returns in the more advanced central European countries than from the more under-developed economies of the Middle East. All of these issues work against the notion that an increase in co-operative levels of interaction between Western Europe and Mediterranean countries is likely. In addition, given the budgetary constraints the EU is already experiencing, it is unrealistic to assume that this international organization could resolve all of the problems confronting the Mediterranean area without the assistance of other such organizations and international powers, particularly the United States.

The more the EU consolidates its position in the Mediterranean, the less likely it is that a Mediterranean international region will emerge. One way that this process may be reversed is if Southern European EU member states adopt foreign policies which are more Mediterranean oriented. If Southern EU member states divert some of their resources towards intensifying intergovernmental and transnational links at all levels with countries along

WESTERN EUROPE

ENLARGEMENT	FREE TRADE AGREEMENTS	NO WIDENING OR DEEPENING
Central	Selective Ties	Integration Halts
East	Marginalization	Lack of Unity
South	*"FORTRESS"*	*"FRAGMENTATION"*
"WIDER"		

Co-operative	Co-operative / Conflictual	Conflictual
[European International Region]	[Mediterranean Frontier]	[Mediterranean Divide / Vacuum]
Co-Prosperity Sphere	3 Constellations (S. Europe - Maghreb - Levant)	Clash of Civilizations (Christian / Muslim)
Outer Zone of Suzerainty		*Economic Wasteland*

MEDITERRANEAN AREA

Figure 7.2 - Potential European-Mediterranean Relations in the Post-Cold War Era

the southern shores of the Mediterranean, then it is possible to envisage a more co-operative dominant pattern of regional dynamics emerging across the Mediterranean. One incentive for Southern European states to follow such a route is the security pay back they would achieve by fostering a more stable southern backyard. However, the advantages of maintaining a northern bias in their foreign policy approaches far outweigh a southern shift. Concentrating their already limited resources on nurturing closer ties in the Mediterranean area would expose Southern European countries to the risk of being marginalized as their northern neighbours focused their attention on a more co-operative Eastern Europe. Southern European states are also aware that any concerted effort they could muster would not compensate for the large disparities which exist in the Mediterranean area. Unlike the co-operative nature of most Central and Eastern European governments, Southern European states also know that several states across the Mediterranean may soon be controlled by Islamic purists who would rather establish stronger links with like minded governments than with Western countries. This cluster of factors makes it unrealistic to expect Southern European countries to align their foreign policies more with the Middle Eastern subregions of the Maghreb and the Levant. By extension, it also means that the concept of a co-operative Mediterranean international region emerging from any European led initiative is a fictitious aspiration.

If the West European international region and its Southern Europe subregion are not serious about assisting in the nurturing of regional dynamics across the Mediterranean which would lead to the creation of a co-operative Mediterranean international region, why is the EU engaged in the area and why do Southern Mediterranean states and their Mediterranean neighbours, particularly those in the Maghreb, pay lip service to the notion of such a Mediterranean international region? Numerous reasons help account for such foreign policy positions. First, adopting proactive foreign policies assists countries in Southern Europe, the Maghreb and the Levant subregions pressurise other states in their respective international regions to upgrade their relative positions within their region. Southern European countries are particularly aware that the post-Cold War security debate has largely focused on a "new" trans-Atlantic security architecture. In an effort to correct this institutional bias, the Southern Europeans have proposed schemes such as the CSCM, "5+5" forum and the Black Sea Co-operation Council (Snyder, 1993: 107). Second, political rhetoric costs nothing and if it can act as a catalyst for transnational patterns of interaction to commence or at least succeed in delaying conflictual patterns of relations from developing, it will have proved its worth. Third, by dedicating its economic and political instruments to the Mediterranean area, the EU could enhance further its sphere of influence and ensure that the Mediterranean does not become a front line in North-South power politics. In the past European responses to challenges in the Mediterranean basin have been characterized by conceptual fragmentation and weak instruments. This lack of EU strategic leadership is a principal reason for the failure to defuse economic and social tensions in the Mediterranean and contributes to the "conflict of civilizations" scenario that some international relations scholars have forecast (Huntington, 1993).

The transition to a new order after the Cold War has altered conflict constellations in every corner of the world, including the Mediterranean. For example, the lifting of superpower overlay in the Arab-Israeli crisis has dramatically changed the dynamics of interaction in the Levant in a relatively short period of time. The disappearance of the Soviet Union has led countries like Syria to search for compromise with its neighbours after decades of an isolationist foreign policy. The defeat of Iraq in the Gulf War enabled the "international community" in general, and the United States in particular, to inject momentum into the peace process that resulted in the 1993 Israeli-Palestinian breakthrough and the 1994 Israeli-Jordanian peace treaty. This process of regionalization at a sub-regional level opens up new prospects for enhancing co-operation at a trans-Mediterranean level. Construction of an elaborate infrastructure in the Levant and the Maghreb would help in the task of furthering intergovernmental and transnational relations in an area where they are weak or absent. As the wealthiest international region in the vicinity, Western Europe will have to play some role in this process if it is to stand any realistic chance of continuing. Recent economic contributions to countries in the Levant to help build a trans-Levant highway is one example of the kind of assistance that could help spur intraregional patterns of interaction in this area. At the end of 1989, Italian Foreign Minister, Gianni De Michelis, did in fact suggest that EC countries should commit themselves to transferring 1% of their GNP in official development assistance by allocating 0.50% to the less developed countries, 0.25% to the Eastern European countries and 0.25% to the Mediterranean countries (Aliboni, 1991: 31). In reality, the EU currently allocates twice as much funding to East and Central Europe as it does to the Mediterranean (Financial Times, 15-11-94: 18). Moreover, the 4.7 billion ECU the EU is dedicating to the Mediterranean for the 1995-1999 period is less than half the trade surplus it registered with the same area during 1993 (12.1 billion ECU) and a little more than half of the surplus obtained during 1994 (9.3 billion) (Eurostat, 1995).

The apparent incoherence and unevenness of EU policy in the Mediterranean during the last two decades is understandable when one considers the disjointed nature of the area. The diversity of the Mediterranean and the challenges it presents makes it politically and economically unrealistic to assume that the EU could introduce and implement a comprehensive programme that would include all the countries in the basin.

7.2.2 The Role of the Organisation on Security and Co-operation in Europe (OSCE)

Another international organization with influence in the Mediterranean is the Organisation on Security and Co-operation in Europe (OSCE; formerly the Conference on Security and Co-operation in Europe, the CSCE). Since the collapse of communism, the 53 members of the OSCE, which groups all European countries with the United States and Canada, has sought to find a new active and dynamic international role for itself. The OSCE is currently actively engaged in monitoring conflicts and ensuring human rights in the Commonwealth of Independent States (CIS). Yet the mire of Bosnia illustrates the limited success the forum has had in the fields of post-Cold War peace-keeping and peace-making. Now the OSCE finds itself in competition with NATO, the EU, the WEU and the United Nations, who are all attempting to develop new mechanisms to contend with the multifaceted security challenges of the nineties.

Critics of the OSCE argue that this international organization has become little more than a flexible talking shop on human rights and minority issues. Proponents of the forum counter this viewpoint by blaming the organization's lack of success on the reluctance of Western governments to back the organization both diplomatically and financially. Without the necessary mechanisms or the military structure, the OSCE cannot fulfil its 1992 Helsinki Summit pledge to become the leader in preventive diplomacy.

OSCE decisions are made on the basis of consensus and can thus be blocked by one dissenting voice. The notion of "consensus minus one", which was used in the League of Nations, was introduced to curb the Yugoslav government's obstructionism. The OSCE has however, no way of compelling participating states to live up to their commitments. Some observers claim that the fundamental principle of consensus is the main weakness of the system, because it is practically impossible to get 53 countries to agree on any given subject. On the other hand, others argue that the principle of consensus within the OSCE is really a strength because it means that all European states are included on an equal basis in the decision-making process. A great deal of the criticism of the OSCE stems from the basic misunderstanding that the OSCE is a co-operative rather than a collective security forum and as such has very limited compellent, deterrent or enforcement mechanisms (Kemp, 1994: 183-5).

As a co-operative security organization the OSCE seeks to stimulate patterns of relations among its participating states that other collective security forums cannot. For example, through its process of discourse and dialogue the OSCE is attempting to structure early warning systems that will help in the field of conflict prevention, assist participating states in fulfilling their OSCE obligations, and helping find compromises when member states have disagreements. During the Cold War, intergovernmental approaches were often supplemented by transnational contacts (cultural exchanges) which helped correct some of the misperceptions which existed on both sides of the East-West divide. This preparatory groundwork has also facilitated the increase in the pattern of relations since the collapse of the Berlin Wall, between Eastern and Western Europe by providing the foundations upon which an intergovernmental and perhaps in time, a transnational European international region, stretching from the Atlantic to the Urals, can be established.

At the December 1994 OSCE gathering in Budapest, the only great power to push the case for making the OSCE a major body for security in Europe was Russia. Moscow proposed establishing a UN-style security council within the OSCE that would have the power to veto actions by other European bodies such as NATO. Russia argues that a genuine division of labour must be set down between the Commonwealth of Independent States (CIS), the North Atlantic Co-operation Council (NACC), the EU, the Council of Europe, NATO, and the WEU, with the OSCE playing a co-ordinating role for the sake of burden-sharing, efficiency and better management of resources.

Western powers do not agree with Moscow's suggestion that the OSCE become Europe's main security forum largely because they are unwilling to subordinate NATO, and their freedom of action within this alliance, to the OSCE. By creating a hierarchy with the OSCE at the top, Russia would retain a powerful voice (veto) on European security issues. Moscow has complained on several occasions that the West is marginalizing it in the post-Cold War security decision-making process and it wants to play a more active peacekeeping role, especially in its "near abroad" - former Soviet territory outside Russia since the Soviet Union's break-up.

With no fixed abode for most of its life, the OSCE now has small offices in Prague, Warsaw, and the Hague as well as a headquarters in Vienna. But it has virtually no permanent staff and a minuscule budget which amounted to $18 million in 1993. In their conclusions at the Third Meeting of the OSCE Council in Stockholm in 1992, ministers confirmed that they would further improve the operational capacity of the OSCE agreed in Paris and Helsinki. But they would simultaneously try to achieve this objective without the creation of a bureaucracy (OSCE/3-C/Dec. 1992: 19). Despite this development, the OSCE has managed to dispatch small monitoring missions to several "hotspots" in the last few years which include: the Serbian province of Kosovo, the former Yugoslav republic of Macedonia, Estonia, Latvia, Georgia and Nagorno-Karabakh (ibid.: 1-12). A restructuring of the instruments at its disposal and an increase in its funding could therefore allow the OSCE to apply its conflict prevention and peace-building experience in other volatile areas in the vicinity of Europe, such as the Mediterranean.

Throughout the Helsinki Consultations in 1973, Malta strove single-handedly to ensure that the principle, whereby security and co-operation in Europe are linked to security and co-operation in the Mediterranean, is enshrined in the Helsinki Final Act of 1975. In the post-Cold War age where the definition of security has evolved to embrace socioeconomic and environmental aspects in addition to traditional political/military issues, the linkage between Euro-Mediterranean security is accepted by most European security institutions, including the WEU and NATO.

Within a Mediterranean context, the OSCE is limited in influence because its membership does not include the majority of actors in the basin. Although the OSCE avoids mentioning specific Mediterranean security proposals, such as the CSCM in the Helsinki Declaration of July 1992 entitled *The Challenges of Peace*, it nevertheless recognizes the significance of the Mediterranean to the future of European security:

> We recognise that the changes which have taken place in Europe are relevant to the Mediterranean region and that, conversely, economic, social, political and security developments in the region have a direct bearing on Europe (OSCE Helsinki Declaration, July 1992, point 37).

The OSCE also commits itself to the widening of co-operation and increasing its dialogue with the non-participating Mediterranean states in an effort to

> promote social and economic development, thereby enhancing stability in the region, in order to narrow the prosperity gap between Europe and its

Mediterranean neighbours and protect the Mediterranean ecosystems. We stress the importance of intra-Mediterranean relations and the need for increased co-operation within the region (ibid., points 39-40. See also Helsinki II Declaration, Chapter IV and Chapter X.).

In spite of the indifference manifested by the OSCE towards the Spanish-Italian CSCM project, it organized a CSCM-type meeting in Malta in May 1993. The OSCE Seminar was held in accordance with the provisions of the Helsinki II Document of 1992. As was the case at Malaga, the Seminar's participants were free to discuss problems pertaining to the Mediterranean without the inhibitions normally associated with the onus of a negotiated text.

The Seminar debate revealed that a significant number of Mediterranean non-European countries maintain a marked interest in keeping links with Europe active within the framework of the OSCE. Requests for more regular consultations were put forward and an interest was expressed in being associated in a more permanent, structured relationship.

Deliberations focused on the political, social, economic, and humanitarian factors behind demographic trends and migration. A call for accelerated economic co-operation was also suggested. This request emphasized that European markets should be further opened to exports of manufactured commodities from the Mediterranean. It also asked for an increase in financial commitments for projects and programmes in the field of agriculture and more support for private sector operations through the encouragement of joint ventures between European companies and non-participating Mediterranean states. Support of macroeconomic and sectoral reforms in non-member Mediterranean countries was also solicited.

When discussing environmental issues, the delegations reaffirmed that all countries in the isthmus should adopt national strategies for sustainable development in accordance with the recommendations of Agenda 21 of the United Nations Conference on Environment and Development (UNCED), held in Rio de Janeiro. Regional co-operation and co-ordination was also encouraged. Actions which were debated comprised:

* establishing and implementing the legal instruments of environmental impact assessment
* strengthening national institutional capacity to integrate social, economic, developmental and environmental issues at all levels of decision-making and implementation

* promoting the rapid introduction of solar and wind energy
* establishing instruments with a view of integrating environmental costs in all economic activities and ensuring that prices reflect the relative scarcity and total value of resources.

Consensus was also achieved on the necessity to review and reinforce a full range of legal instruments relevant to the Mediterranean area. Particular emphasis was directed toward those instruments aimed at protecting and restoring Mediterranean ecosystems as identified in the Mediterranean Action Plan and the Barcelona Convention. Other legal instruments, such as the convention on Climate and on Biodiversity, the Barcelona Convention on a global level and the ECE Conventions on a European level, were also recommended for ratification. The elaboration of new conventions and protocols such as the Convention on desertification were also presented. Delegates at this OSCE Seminar stressed the priority that should be attached in international co-operation to cross-sectoral requirements such as institution-building, development of endogenous capacities and technology transfer (Calleya, 1994b: 30-40).

At the December 1994 OSCE Summit in Budapest, Mediterranean member states, such as Italy and Malta, advanced the case that OSCE "associate member" status could be created for non-participating Mediterranean countries. Such membership would be open to those non-member Mediterranean states that are prepared to abide by OSCE principles. Offering states such as Egypt and Tunisia associate partner status opens up avenues for closer political co-operation and gives these states the opportunity to become acquainted with the OSCE experience in early warning exercises, conflict prevention, and crisis management techniques. As such, the introduction of associate membership to non-participating Mediterranean states would promote intergovernmental social and political and military (preventive diplomacy) types of interaction across the waterway.

Such a development would certainly have an immense impact on the nature and scope of regional dynamics in the Mediterranean area. Any OSCE led effort to increase intergovernmental and transnational links between Europe and the Middle East subregions of the Levant and the Maghreb would have the advantage of involving all of the leading Western great powers and Russia. Although consensus among such a diverse group would be difficult to reach, it would also guarantee that all those actors with an interest in the basin could participate.

If this was to happen the 53 members of the OSCE would have to remodel their forum which was tailor made to contend with the East-West security challenges of the Cold War. One way that the OSCE could enhance its influence in the Mediterranean is by developing its potential as a 'regional arrangement' under Chapter VIII of the UN Charter (CSCE, 1992: 21-22). A joint Dutch-German initiative advocates a policy of "OSCE first". Participating states would be encouraged to commit themselves "to every effort to achieve pacific settlement of local disputes" through the OSCE before referring them to the UN. Balancing its general principles of being a co-operative security organization with effective enforcement mechanisms would enable the OSCE to become a mandate issuing body. This initiative would create the possibility of launching missions under OSCE auspices which include the supervision of cease-fires, the monitoring of troop withdrawals, the respect of the rule of law and order, and the provision of humanitarian assistance to refugee groups (Calleya, 1994b: 30-40).

If the OSCE were to extend its membership (even associate) to non-participating Mediterranean states who agree to abide by the forum's regulations, it could act as a catalyst towards enhancing more intense co-operative regional dynamics across the Mediterranean basin. Inclusion of interested countries from the Mediterranean area will assist in ending the perception that the OSCE is yet another great power concerted attempt to dominate and preserve the status quo amid the weaker countries in the area. In its favour, the OSCE has the advantage that it is an inclusive or co-operative security institution. By its very nature it seeks to avoid punitive or forceful actions against participating states. In an area as diverse as the Mediterranean, this approach would allow all the actors concerned to debate sensitive issues that cannot be discussed in other intergovernmental forums. The OSCE Mediterranean Seminar held in Tel Aviv in June 1996 went some way to commencing such a process by further promoting the concept of the OSCE as a platform for dialogue and fostering norms of behaviour across the Mediterranean.

As stated above, the OSCE can apply mechanisms it employed during the Cold War to promote East-West transnational types of relations long before they were regarded possible, in the Mediterranean area, where they are currently regarded as unfeasible. Yet, laying such foundations has to be seen as a long-term endeavour. After all, the East-West rapprochement process started by the CSCE in Europe in the 1970s is only now bearing tangible results and manifesting itself through programmes such as NATO's NACC and PFP initiatives and the EU's Central and Eastern European enlargement programme. Just as evolution of the notion of a European comprehensive international region has taken decades to develop and has evolved in stages, i.e., intergovernmental level followed by an increase in transnational interaction, the evolution of a similar process in the Mediterranean must be deemed as a long-term goal given the political, economic, social and military disparities which have to be harmonized across the Mediterranean.

7.2.2.1 The OSCE and the CSCM: What Relationship?

While the OSCE has made no direct reference to the concept of a CSCM, it recognizes that European and Mediterranean security are interdependent (Ghebali, 1993: 97). An attempt to transpose the Helsinki model for the OSCE to fit the Mediterranean raises the question of what nuances and safeguards such an endeavour would entail. One significant difference in this regard is the nature of the situations the OSCE and the CSCM have to operate in. During the Cold War, antagonisms with Central and Eastern Europe related mainly to ideological, socioeconomic, and military issues. In the case of the Mediterranean, the socioeconomic factor would be coupled with cultural disparities (Yahia, 1993: 7). Thus, while a CSCM process would benefit enormously from imitating the OSCE experience in the areas of economic development, human rights, and resolution of conflicts, it offers little assistance on how to overcome the cultural divide which exists in the Mediterranean area. While the OSCE was erected in a rather stable atmosphere - the German Democratic Republic had been recognized, the Soviets had offered concessions over Berlin - a CSCM would have to be established on a much more unstable terrain, due to the ongoing stalemates in Cyprus, the Western Sahara and the Arab-Israeli conflict.

In 1990, proponents of the Spanish-Italian CSCM proposal regarded it as an attempt to extend the OSCE-type mechanisms to an area which lacked the benefits of a single institutional framework. Southern European countries also used the CSCM initiative to counter-balance the increase in attention Western European states were beginning to dedicate to Central and Eastern Europe. The Gulf War of 1990-91 had already shifted international attention to the Mediterranean area. Realizing that international security concerns in the vicinity would quickly wane once a cessation of hostilities came about, Italy and Spain launched the CSCM scheme in the hope that they could harness international support behind a co-operative security initiative for the area. It is also possible that the CSCM initiative was launched to attract the OSCE to Mediterranean security issues. Rather than face competition from another security forum, Southern European states, and especially non-OSCE member Mediterranean states, may have calculated that a CSCM proposal would impel the OSCE to concentrate more of its resources to the Mediterranean area.

A common problem that the CSCM would share with its European counterpart, is that of obtaining United States support. Washington already participates in a less important role than it is accustomed to in the OSCE, and it seems unlikely that the U.S. would forfeit its freedom of movement in the Mediterranean area to become part of a complex and inexperienced security arrangement. In addition, a CSCM would also face the difficulty which is plaguing most security institutions in the 1990s, namely, how to implement agreed upon resolutions. Without access to U.S. military technology it is, for example, difficult to perceive how a CSCM could monitor non-proliferation regulations. Lastly, setting up a CSCM would add to the multiplicity of already existing security institutions in the vicinity. Rather than duplicate such organizations it may seem more feasible to invest already scarce resources in strengthening bilateral security ties between well established international organizations such as the OSCE, NATO and WEU and non-member states in the Mediterranean (see Rato, 1994: 13-15).

Even if a CSCM were to be established it would still have to overcome the daunting challenges that other international organizations are having to confront. For example, would a CSCM limit itself to strictly humanitarian programmes or would it also undertake conflict-prevention missions? Where would the finances required to operate such a large international forum be found? What voting mechanisms would be adopted to ensure that a CSCM is both effective and accountable for its actions? And finally, where would a CSCM fit in the hierarchy of already existing security clubs? Would it be a junior OSCE or would it be an autonomous security organization with an independent mandating authority? As is evident from the above, the notion of establishing a CSCM raises as many questions as it potentially provides answers. Until already functioning international organizations can work out what roles they are to play in a post-Cold War security framework it seems unlikely that additional multilateral organizations, such as the CSCM, will find the necessary political or economic support to become a reality.

7.2.3 The Role of the North Atlantic Treaty Organization (NATO)

In the last few years the North Atlantic Treaty Organization (NATO) has been going through a period of soul-searching in an attempt to identify what new roles it can play in the post-Cold War security agenda. In the Mediterranean, the Atlantic Alliance has sought to extend its multilateral approach in two ways: by sending confidence-building signals to non-member Mediterranean states and by fostering a politico-security culture similar to that which exists in Western Europe.

Traditionally, NATO has always included the Mediterranean dimension in its forecasting: Italy was among the twelve original signatories of the Treaty. After 1949, the Alliance reaffirmed its commitment to this area as three of the four new members were from Southern Europe. The significance attached to the Mediterranean and its flanking areas is further highlighted by the fact that two-thirds of the some fifteen Alliance relevant regional conflicts have occurred in this area since 1956 (George, 1993: 9).

NATO's approach to the Mediterranean has undergone considerable evolution since 1949. Three specific innovations took place during the Cold War which retain their significance today:

* A special group for consultations on the Mediterranean was established in 1967 in an application of Article IV of the Washington Treaty. This consultative process promotes consensual views among the Alliance members.
* In 1975 the Alliance endorsed the idea of a Mediterranean basket within the framework of the OSCE. This step is one of the first to recognize the importance of an institutional dimension to trans-Mediterranean affairs.
* In 1982 the Alliance confirmed that it was legitimate for member states to respond unilaterally to requests from third parties for assistance against aggression and that such actions would be compensated for by other NATO assets. As a result, redeployments from the Mediterranean were often replaced by those from other member states.

Throughout the Cold War both the Canadians and the Americans co-operated with their European allies to ensure strategic depth and deterrence in the geographically and geopolitically varied parameters of the Mediterranean. David Law, former Head of the NATO Policy Planning Section in the Political Directorate identifies the following accomplishments by the Alliance in the basin during the Cold War:

* it ensured strategic cohesion across Southern Europe;
* it provided security insurance that was respected;

* it advanced stability and political rapprochement between countries traditionally at odds with one another;
* it maintained the defence modernization of Southern European members (Law, 1993: 5).

NATO's strategic stand-off in Europe indirectly enhanced security in the southern periphery of the Mediterranean by containing the hostilities that did break out from time to time.

In the post-Cold War world of the 1990s, NATO has attempted to transform its posture and approach to the Mediterranean. The first change is conceptual in nature. At the 1989 Anniversary Summit and in the Strategic Concept document of 1991 in Rome, the Alliance acknowledged the multifaceted security challenges confronting it. In addition to military threats, NATO also recognized problems such as weapons proliferation, terrorism, economic disparities, environmental degradation and mass migration. NATO also underlined its acceptance to participate in an interlocking institutional arrangement in which different security institutions, i.e. the OSCE, WEU and UN all contributed to the resolution of conflicts.

The second change that has taken place relates to NATO's defence reorganization. The Alliance is seeking to develop its operational mobility and flexibility. An emphasis is also being put on increasing multinational operations. It is hoped that such developments will equip the Alliance to address the spectrum of challenges in the Mediterranean.

NATO's third shift is linked to the practical experience the Alliance has had in the Mediterranean area over the last few years. Three particular episodes stand out. First is NATO's contribution to the 1991 Persian Gulf War which is paradoxical in nature. Although not directly involved in the Gulf itself, NATO played a decisive role in the successful military prosecution of the war. Consultations, information-sharing and policy concertation, conducted through Alliance channels, helped to galvanize international support for UN resolutions against Iraq. NATO's deployment of its mobile air force to Turkey not only bolstered the latter's defences, but also enhanced intelligence collecting throughout the Mediterranean. The Alliance also supplied essential logistical and communications support to the twelve nations who actually had forces on the ground during the war (Snyder, 1993: 102-19). The 1991 Gulf War advanced the thesis that post-Cold War crises could be dealt with successfully if the indispensable importance of harmonizing the Washington Treaty with the UN Charter was recognized.

This thesis was underlined by NATO's role in a second episode it had to face, the Balkan crisis. The lack of a concerted North American/European perspective on the military and political demands associated with the conflict in the former Yugoslavia is one reason that UN resolutions in the area have not been fully enforced. As the war has continued, NATO has attempted to contain the conflict by furthering its activities in a number of ways: participating in monitoring operations in the Adriatic alongside the WEU under UN resolutions; overseeing UN authorized no-fly-zones in Bosnia; increasing collaboration between the UN, the WEU, NATO, and co-operation partners such as Hungary and Albania; and through individual member state participation in UN peacekeeping missions.

The third episode that has affected NATO in the 1990s is the Middle East Peace Process. Although not directly involved in the regional dynamics of the Levant, NATO is a decisive third party that acts as a buttress to the functional EU economic role and US military role in the area. This fact was clearly demonstrated in the 1990-91 Gulf War when NATO refuelling bases and other logistical support greatly accelerated the Desert Shield build-up and the Desert Storm campaign (Snyder, 1993: 109-10). NATO's involvement in the Gulf War and its more recent participation in the Balkan crisis set precedents which make it highly unlikely that the "out of area" issue will prohibit this international organization from intervening in future crisis situations across the Mediterranean. This development in NATO policy presents both risks and opportunities for countries in the basin. A violent collapse of authority in states such as Algeria could easily prompt a rapid reaction NATO response to safeguard expatriates and protect vital petrochemical installations. Conversely, countries in the Mediterranean can take advantage of NATO's renewed interest in the area by opening wide-ranging security discussions with the Alliance. Such a process of dialogue could help dispel some of the misperceptions which exist on both sides of the Mediterranean.

NATO's institutional efforts to further ties with its northern and eastern neighbours through the North Atlantic Co-operation Council (NACC) and the Partnership For Peace (PFP) programmes have more recently been emulated in the Mediterranean space of operations. For example, in April 1995 Malta became the twenty-sixth member of the PFP programme, thus extending this network of NATO to the heart of the Mediterranean. NATO has also outlined plans to establish contacts, on a case-by-case basis, between the Alliance and non-NATO Mediterranean countries (Claes, 1995: 7). Details of a NATO "southern strategy" were spelled out by NATO officials in February 1995. Five countries in the Maghreb and the Levant (namely Morocco,

Tunisia, Mauritania, Egypt, and Israel), have been selected as the first countries in the Middle East international region which NATO is prepared to open a dialogue with (Financial Times, 9-2-95: 6). Although there is no mechanism to co-ordinate NATO's "new" Mediterranean policy with that of the EU, the two international organizations are co-operating through the fledgling EU defence arm, the WEU, in the formulation of this policy which aims to contain sources of instability in the Mediterranean area (ibid.). At the 1993 Annual Session in Copenhagen, the North Atlantic Assembly unanimously urged NATO governments to launch an "active outreach programme directed towards the Southern Mediterranean basin, which would assist in facilitating dialogue", (Rato, 1994: 15). Former NATO Secretary General, Manfred Woerner, did not respond to this particular point, perhaps indicating the delicacy of this issue within the Alliance. NATO currently insists that it has no intention of setting up any new organizations in the south. The introduction of a "NACC-South" has however been discussed as a logical way for the Alliance to advance its presence in the Mediterranean and thus become a dominant multilateral intrusive actor (this issue has been raised at the four so-called Rose-Roth Seminars on Security and Co-operation in the Mediterranean in Athens, Granada, Istanbul, and Capri). A number of indicators support the creation of such a forum:

* NATO's advanced command and control structure could serve as a vehicle for promoting security dialogue with non-member countries;
* the lessons learned through the outreach programme with Central and Eastern Europe are applicable in the South as the security challenges in this area are also related to transitory post-Cold War realities;
* NATO's links with Central and Eastern Europe and Central Asia through NACC provide a forum through which ties between these regions and the Mediterranean could be developed;
* by extending its diplomatic and military machinery southwards now, NATO, together with other European organizations like the WEU, the Council of Europe and the EU, could ensure their involvement in any future trans-Mediterranean security initiatives, such as the proposed CSCM.

NATO's recent initiatives in the Mediterranean reflect a genuine effort to strengthen its influence in an area where other organizations of a different ethnic background such as the Arab Maghreb Union, the Arab League, and the Black Sea Co-operation, already operate. NATO's increasing concern with developments in the Mediterranean also reiterates the Alliance's 1994 Declaration that security in Europe is greatly affected by security in the Mediterranean (Declaration of the Brussels NATO Summit 10/11-1-94).

The post-Cold War period is proving to be a revolutionary era due to the fact that dividing lines of the past have faded or disappeared completely. Yet no clear pattern of international relations has emerged in their place. This period of rapid flux presents NATO with an identity crisis which is exacerbated when seen through the lens of such a diverse area as the Mediterranean. But it also presents the Alliance with an opportunity to forge new links with Mediterranean non-members. Although the costs of developing an active alliance network across this waterway will be high, the costs of failing to establish such a system could be higher in the long-term, should instability from the Mediterranean spread northwards.

NATO's successful Cold War track record makes it one of the most prominent security institutions functioning today. One way to preserve this position is by leading co-operative efforts with other institutional associations that also have an interest in ensuring stability in areas like the Mediterranean. By forming coalitions and relationships with other international organizations in the basin, NATO could play a direct role in helping to prevent the emergence of conflictual patterns of relations between the Western European and Middle Eastern international regions. Such relations could easily evolve if political and military misperceptions and the increase in the proliferation of weapons is not checked in the short-term. Participation in NATO activities will assist in removing some of the negative perceptions non-member Mediterranean countries harbour about the Alliance. For example, permitting countries in the Maghreb and the Levant to attend certain NATO sessions will illustrate that NATO is a common defence grouping and not an aggressive military alliance. It has also been suggested that those Southern Mediterranean nations who are emerging as democracies should be afforded observer status in NATO (Rato, 1994: 17). The inclusion of non-member Mediterranean countries in NATO's consultative framework would help remove existing misperceptions on both sides of the Mediterranean basin and could help generate co-operative intergovernmental interaction in the sensitive area of military issues.

As the most active military international organization in the Mediterranean basin, NATO has the capacity to influence the patterns of relations across and around the Mediterranean basin. In the end, it will come down to the Alliance's ability to read the indigenous patterns of relations and act according to these trends. Like other international organizations operating in the Mediterranean, NATO is finding it difficult to implement a comprehensive and coherent security programme in this area. Rather than solely blame the Alliance for this outcome it seems more accurate to indicate that NATO's apathy in the South is more the combination of the Mediterranean's incoherent and diverse regional dynamics and NATO's inability to act.

7.3 Conclusions

The emergence of a more multipolar international system in the last five years has seen an increase in multilateral intrusive behaviour in the world. Great powers are eager, to at least appear, to be acting multilaterally in their foreign policy endeavours. The United States' emphasis on obtaining a UN mandate before it acts outside its borders as it did in the Persian Gulf War in 1990-91, Somalia in 1992-93 and Haiti in 1994 illustrates this trend.

The concluding section of this chapter attempts to identify the impact international organizations have on regional dynamics in the Mediterranean area. The end of the Cold War, the process of European Union and the winds of peace blowing from the Middle East have changed the parameters of Mediterranean regional politics. One significant shift is that Mediterranean littoral states are much more keen to develop active relations with the rest of the world. Although they are still apprehensive about the implications of an enhanced American or European role in the area, they actively seek relations with the West now that competition for foreign direct investment has increased.

A United States security presence in the vicinity as a balance against the revival of old or new hegemonic threats, or new terrorist threats under the guise of Islamic fundamentalism, is also still favoured among the majority of countries in the basin. Appeals to establish a nuclear free weapons zone (Palma, 1992) or to establish a multilateral security forum may find their place in the future, but so far both remain symbolic aspirations.

Post-Cold War considerations have led outside powers and international organizations to re-evaluate their policies towards the Mediterranean. Conversely, regional leaders have had to explore new external alignments in light of the sea-change in the international system since 1989. Two recent changes in the dynamics of the Mediterranean regional politics may affect the nature of intrusive influence in the Mediterranean. First is the Arab-Israeli reconciliation process. The peace treaties signed between Israel and the PLO and Israel and Jordan may become preliminary steps towards establishing a co-operative pattern of intergovernmental relations in the Levant. Rapprochement between Israel and the Arab countries has the additional benefit of removing one of the stumbling blocks that has prevented closer relations between the Levant and other Middle Eastern subregions such as the Maghreb. The Middle East international financial meeting held in Casablanca at the end of 1994 highlights the potential that peaceful relations can bring to this international region. If stability persists, and this is no foregone conclusion, attracting foreign direct investment to the area may become a more feasible enterprise. Such a development would fit in with the statement made at the start of this chapter, that nonregional actors are most influential in international regional relations when they complement the basic pattern of regional alignment and conflict.

The second shift in Mediterranean politics is both internally and externally motivated. After years of being accused of marginalizing and isolating its southern flank, the EU has proposed establishing a free trade zone in the Mediterranean area, incorporating the Levant and the Maghreb. At first, it might appear that the EU is on the path of establishing some kind of co-prosperity sphere or an outer zone of suzerainty in the Mediterranean. In reality, it must be stressed that the EU has yet to formulate a coherent vision of long term goals that could justify such possibilities. In addition, the EU does not have the necessary resources to contend with the disparities which exist across the Mediterranean. At most, the EU's most recent free trade proposal can promote economic interaction among states in the area and indirectly promote political exchanges. It is unlikely that the EU's free trade plan will elevate or even address cultural, environmental, or military relations.

If successful, this extra-Mediterranean led effort to enhance economic co-operation between Europe and the Middle East could indirectly benefit trans-Mediterranean initiatives. The evolution of an interdependent Mediterranean economy would make it more difficult for actors in the basin to upset the balance of power due to the economic consequences they would have to confront as a result. Economic interdependence could eventually lead these riparian states to unite on certain issues and to put additional pressure on their extra-regional principal trading partners.

The fact that Western Europe has developed a multi-level international society in which international organizations such as the European Union, the OSCE, and NATO, can interact with states and sub-national institutions, puts this comprehensive international region in a strong position to approach security issues in the Mediterranean in a multi-institutional and multi-functional manner. Given the lack of unity in the perceptions of the countries in the Mediterranean and those powers with an interest in the area, it is unrealistic to assume that a single international organization can contend with the security challenges across the Mediterranean. A more realistic alternative is one in which a single international organization, for example the EU, is assisted by others who have an interest in the area. As the international organization with the largest proportion of Mediterranean member states and the most active socioeconomic actor in the Mediterranean, the EU seems the best positioned candidate to lead European initiatives in the south. In comparison to other organizations with an interest in the area, the EU is perceived in a much more positive fashion by the majority of countries in the Mediterranean. For example, NATO's Cold War military record makes it an unattractive partner to several countries in the Middle East. An increased NATO presence in the Mediterranean could even increase accusations of "neo-imperialist" designs by Arabists and Islamists and thus fuel support for the already very active Islamic fundamentalist groupings operating throughout the Middle East region. American participation in both NATO and the OSCE also makes both organizations appear more like vehicles of great power interests than ones concerned with advancing Mediterranean causes. Absent the creation of a trans-Mediterranean international forum, which would certainly be perceived as much more representative of Mediterranean regional interests and not some self-referenced or great power interests, the EU appears the most acceptable international organization across the Mediterranean that can intensify co-operative patterns of relations throughout this area (Calleya, 1996: 43-54).

Post-Cold War international events show that multilateralism has failed to address effectively the increase in domestic regional hostilities. Over the last five years most regions of the world have been touched by a resurgence of such intolerance based on traits that include ethnicity, language and religion. The Mediterranean space is no exception. The ad hoc and often ineffective international response to many of these crises has cast a question mark on the relevance of the multilateral mechanisms designed to contend with different types of problems, i.e., of an international nature. Civil conflict and regional tensions are not the only security issues that need to be addressed in the Mediterranean. Yet, international organizations must adapt their *modus operandi* if they are to play a pivotal role in diffusing such contentious issues as environmental degradation, economic disparities, migration, weapons proliferation and narcotics trafficking (Anyaoku, 1994: 318).

The United Nations remains the principal international organization for achieving such multilateral endeavours. The UN is, however, already suffering from overstretch and cannot be expected to focus on such an extensive list of challenges on its own. Other institutions and agencies in the area such as

NATO, the EU, the WEU, the OSCE, the Arab League and the AMU will also have to play a supportive role to the UN if an effective multilateral Mediterranean mechanism is to emerge. In a world without a political, ideological or geographical straitjacket, each institution or agency can play on its comparative advantages to ensure maximum effectiveness.

Multilateral agencies must however be cautioned against expecting rewards from their efforts in the short to medium term. In an area as diverse as the Mediterranean, regional co-ordination and co-operation is probably the most that can be initially achieved. For example, a more active OSCE in the Mediterranean can lead to an increase in political, social and environmental exchanges. Non-member OSCE states in the Mediterranean have already shown a keen interest in co-operating in this forum and there has also been a call to extend associate membership to this area (CSCE Seminar, 1993 and CSCE statement by Maltese Ambassador, October, 18, 1994). Such multilateral governmental action could lay the groundwork for similar exchanges at a transnational level. If supplemented by non-governmental organizations which are already active in the area, existing disparities between the Western European and Middle Eastern international regions can be gradually bridged.

Several countries bordering the Mediterranean have sought external support to help create a single institutional framework in which discourse and dialogue on Mediterranean issues can take place. On the other hand, states such as Libya and Syria remain reluctant to actively engage themselves in such endeavours for a number of reasons: sometimes because of animosities dating back to former colonial days and also due to mutual rivalry among themselves for spheres of influence. The superpower track record in the Mediterranean offers two cautionary notes in this respect. First, external actors can only influence and not dictate regional dynamics. International organizations such as the EU must therefore read and decode the mixed signals originating in the Mediterranean if they do not want their effort to consolidate a sphere of influence across this waterway to result in a conflict based international region (Financial Times, 20-10-94: 2). If cross-border economic measures are introduced in consultation and agreement with the Mediterranean states, the EU free trade scheme could act as a catalyst toward regional collaboration in other areas.

The Mediterranean's dependence on EU commercial markets is an important factor in this equation. It affects the foreign policy agenda of the littoral countries for whom EU trade is the engine of growth. The majority of Mediterranean states all have trade and investment links which already make them an integral part of the European trading zone (Colbert, 1992: 266). Nevertheless, the EU will have to advance carefully if it is not to upset the concept of 'balancing' in relations between Mediterranean states and their external patrons. If nonregional actions are perceived as attempts to dominate intra-Mediterranean patterns of interaction, the latter could retaliate by uniting and becoming less co-operative in their dealings with external actors who have substantial political and economic interests in the area. This would certainly be the case if such a trans-Mediterranean backlash included the key oil and gas producers.

In the post-Cold War period domestic politics play a major role in foreign policy considerations. This trend is likely to continue as internal interest groups become more assertive. This is especially the case in countries across the Maghreb, particularly Algeria, where Islamic movements are already constraining government policies. If current Arab regimes are not pressured by external actors at both a bilateral and multilateral level to establish working relationships with other political activists within their boundaries, the aspiration of nurturing more intense co-operative patterns of trans-Mediterranean relations will surely recede.

It is a truism that the end of the Cold War has released the superpower grip on the Mediterranean. But the indicators discussed above suggest that one type of intrusive dominant system (bipolar superpower model) has been swept aside only to make room for a different type of intrusive dominant system (multipolar great power model). This more multipolar design is reflected in the increase of activity registered by international organizations in regional relations. The more non-Mediterranean multilateral organizations come to dominate patterns of relations in the Mediterranean area, the more they are likely to stifle a resurgence of intra-Mediterranean patterns of relations. As a result, contemporary European international organization involvement in the Mediterranean is best seen as a boundary management exercise, which aims at safeguarding the regional dynamics of integration in Western Europe from those of fragmentation which are active in the Middle East.

8 General Conclusions:

A Resurgence of Regionalism in the Post-Cold War World? Regional Dynamics in the Mediterranean Area

International regions are characterized by a distinctiveness and proximity, not only in geographical terms, but also in a social, economic, political and military sense. Proximity is institutionalized by means of mutual interaction and common organization.

The Mediterranean basin is an example of a geographical region which does not yet meet the requirements of distinctiveness and proximity for an international region, but can rather be described as a "mosaic of subregional institutions and overlapping narrow-focused international organisations" (see Sneider, and Borthwick, 1983: 1245-54). Thus, while some politicians and certain international relations scholars aspire to a Mediterranean international region, post-Cold War Mediterranean realities portray a picture of fragmentation. What are the forces pushing the Mediterranean from a Cold War to a "cold peace"?

During the ancient and classical era, international regions were the dominant structure in the international system. In the 19th and 20th centuries, the system level was so successful that it practically eliminated the international regional approach altogether. During the Cold War, the Mediterranean area was influenced by global conflict and by intrusive dominance. Throughout the latter half of the twentieth century, a situation between these two historic extremes has emerged. The global interdependent system remains strong. However, the process of decolonization coupled with the end of the Cold War has created an environment which is conducive to a resurgence in regional patterns of interaction. As a result, regionalism is again becoming a major characteristic of the international system (Buzan et al., 1994c: 8).

The relationship between a system and its institutions is dialectical. If co-operative patterns of interaction are dominant each particular regional system provides the rationale for the creation of regional institutions that reinforce the regional system itself. If antagonistic patterns of relations are dominant, regional institutions are unlikely to be able to perform the functions expected of them, leaving the former regional system vulnerable to an implosion, or the possibility of extraregional power intervention (Awad, 1994: 158). The Mediterranean has witnessed both of these outcomes. The elimination of superpower geopolitics and the increasing importance of geoeconomics, has forced all the countries in the Mediterranean to reassess their foreign policy priorities.

Superpower overlay has not yet been replaced by international regional patterns of interaction. The Mediterranean is an area where subregional dynamics are a priority. The analysis of the patterns of relations within and between the subregions encompassing the Mediterranean in chapter four, clearly demonstrates this fact. There it was explained that, for the countries of Southern Europe, bilateral relations with countries in the European Union are a primary concern. Links with great powers such as the United States and Japan figure second. Relations with countries in the Maghreb and the Levant figure further down the foreign policy priority chain, with Mediterranean policies given even less attention. In the Maghreb, bilateral relations with European Union member states and nonregional powers such as the United States and Japan are high on the list of foreign policy priorities. Relations with the other members of the Arab Maghreb Union also come before considerations towards the Mediterranean. In the Levant, bilateral relations with neighbouring countries remain a priority. Relations with the United States are also high on the foreign policy agenda, while links with the European Union and Japan have increased in recent years. The result is that few diplomatic resources are left over to dedicate to Mediterranean security issues.

To date, initiatives to establish a trans-Mediterranean international institution, such as the CSCM, have failed because they have been attempting to institutionalize regional patterns of amity which do not exist. A process of transformation is currently underway in the Mediterranean area but by no means do emergent patterns of relations indicate an intensification of contacts on a pan-Mediterranean level. As indicated in the framework of analysis (2.3), to qualify as an international region, the countries' pattern of relations or interactions must exhibit a particular degree of regularity and intensity to the extent that a change in the foreign policy actions of one of the players will have a direct impact on the neighbouring states' foreign policy. The multiplicity of foreign policy models active in the Mediterranean basin, (defensive, offensive, co-operative, neutral) reveals that no firm co-operative or conflictual regional pattern of interaction exists. Thus, while multilateral co-operative regional endeavours have been suggested, the present picture of fragmentation in the Mediterranean and the fragility of the existing process to implant a Mediterranean identity, tends to make subregional bilateral relations a more feasible channel of communication.

A review of such relations shows that subregional and local systems have become the dominant factor in some areas, as illustrated by developments in the Levant. In others, such as Southern Europe, the tendency is toward being progressively immersed in a broader Western European international region which extends beyond the Mediterranean. Regardless of which of these new regional systems emerges, the notion of a Mediterranean international region does not mirror contemporary regional dynamics in this area. Instead, the Mediterranean space consists of a number of semi-autonomous subregional units: Southern Europe, the Maghreb and the Levant, which are constantly becoming more embroiled in their own international regions, namely Western Europe and the Middle East. The Mediterranean sea is therefore more of a frontier. As disparities between the North and South multiply, the area is rapidly becoming a faultline between two separate and increasingly polarized regions (Cerretti, 1993: 4). While regional dynamics of integration are active in Western Europe, patterns of fragmentation continue to dominate regional relations in the Middle East.

This observation casts a cold shower of reality on Mediterraneo-centricists, who perceive a Mediterranean region "decoupling" itself from the dominance of the great international capitalist powers, principally the United States, and forecast that the Mediterranean sea will become kind of a "co-prosperity sphere" between Southern Europe and the Arab countries. They aspire that the Mediterranean will become an international region, that is to say, an area of central focus in its own right. The CSCM proposal exemplifies such a perspective.

In contrast, it reinforces the view of other international observers who perceive that the Mediterranean basin has become a frontier or a boundary, around which many subregional formations gravitate. The long list of internal and external factors functioning in and around the Mediterranean in the post-Cold War is more akin to a frontier between a mosaic of subregional constellations than any particular type of international region itself. The three subregions encompassing the Mediterranean can be classified as follows:

i) a Southern European subregion (Southern European EU member states including the EU applicants of Malta, Cyprus and Turkey), which prophesises partnership with Mediterranean countries, but often portrays signs of a "fortress Europe". This area has strong intergovernmental and transnational features and is the subset in the Mediterranean area which has the most comprehensive links with the rest of Western Europe. Participation in any future steps to establish a single European currency or the framing of a common EU foreign and security policy will guarantee that this subgrouping remains firmly embedded in the Western European comprehensive international region.

ii) a North African subregion (of which Egypt is equally a part) which is seeking to shape closer economic, and perhaps, political, ties with the European Union. Transnational ties are limited to the areas of energy, Islam and ethnicity, with comprehensive international regional features non-existent in this area. Despite political rhetoric claiming the contrary, intergovernmental ties are conflictual dominant in this subregion. For example, the Sub-Saharan conflict between Algeria and Morocco, border clashes between Libya and Tunisia and the more recent (August 1994) diplomatic expulsion of Algerians from Morocco illustrate such trends.

iii) a Levant subregion (in which Turkey is a political, economic, and military player) where intergovernmental co-operative regional links have recently supplemented dominant conflictual patterns of relations. Transnational ties remain at an embryonic level and comprehensive interaction is completely absent. This subregion includes a strong political and military American residence, and a strong European and Japanese economic presence.

A review of the nature of the interactions within each of the subregions bordering the Mediterranean also revealed different evolutionary patterns operating within each subsystem.

The end of the Cold War has had three important geopolitical implications in the Mediterranean. First, the East-West dimension within the Mediterranean has largely disappeared, and Russia's leadership is now committed to establishing firmer links with the West. Second, North Africa has emerged as a "grey area" which is trying to find its position between its Islamic background and the "westernization" process it experienced during the colonial period. Third, the United States's interest in the Mediterranean basin is shifting eastward to the Near East. The U.S. is heavily committed to helping solve the Arab-Israeli division and assist in stabilizing the political equation in the Gulf.

These three developments give further credence to the thesis that the notion of a Mediterranean cross-roads is gradually being replaced by the notion of a Mediterranean frontier whose dynamics are exclusively North-South. Some international security analysts refer to this development as the southern branch of a wider arc of crisis, "the new arc of crisis", which includes all those countries between the Balkans and Russia (see Holmes, 1995: 1-9).

Post-Cold War international geopolitical, and especially geoeconomic, dynamics are reinforcing the pattern of disconnections between the countries in the Mediterranean area and forcing riparian states to look beyond the Mediterranean basin to enhance their security interests.

Across Southern Europe strong intergovernmental and transnational patterns of interaction are present. As part of the Western European international region, this grouping of countries also demonstrate comprehensive regional tendencies by the fact that they are prepared to pool their diplomatic and economic resources to form a European international region. In contrast, no such patterns of interaction are noticeable in either the Maghreb or the Levant. Attempts to emulate the Western European regional experiment have largely failed as demonstrated in the ineffectiveness of both the Arab League and subregional groupings such as the Arab Maghreb Union. In the Middle East relations remain intergovernmental dominant, and are more often than not conflictual. Transnational patterns of interaction across this international region remain limited to joint infrastructural projects, and cross-border Islamic exchanges.

Although a certain level of interaction takes place between Southern Europe, the Levant and the Maghreb, there is no evidence to suggest that the three separate and distinct subzones will emerge as a single international region in the future. In fact, rather than showing any signs of merging, it appears much more likely that one subsystem, Southern Europe, and by extension Western Europe, could come to dominate the others as it continues to establish the parameters of discourse in the area. Current European initiatives in the Mediterranean will not bring about a solution to the existing division, and could lead to the emergence of a new Wall, this time between the northern and southern peoples of the Mediterranean. In such circumstances the Mediterranean would become little more than a *cordon sanitaire* monitored by a European police force (see Ortega, 1995: 49).

The collapse of the Soviet Union and the limited international withdrawal of the United States since the end of the Cold War, has allowed the European Union to emerge as a viable alternative international patron in the Mediterranean. Aware of the economic benefits that enhanced relations with the EU can create, most riparian countries are attempting to upgrade their ties with the EU through bilateral negotiations with the Council of Ministers. Thus, because of the dependence of countries in the Mediterranean on the export of goods and labour to the EU and the importance of stabilizing these countries as a vital European goal, a network of both multilateral and bilateral links between the Western European and Middle Eastern international regions is being reinforced (Hitti, 1994: 87).

The disappearance of communism has also seen a multiplication in the number of actors now competing for foreign direct investment. The Mediterranean countries' arch-rivals are the emerging economies of Central and Eastern Europe. The EU already dedicates more of its assistance budget to this area than to the Mediterranean. An EU enlargement eastwards with no parallel policy southwards would fuel fears of marginalization in the Mediterranean and could spur existing divisions among Mediterranean countries as they try to secure the best deal possible for themselves with alternative patrons.

The nature of the interaction taking place between the three subregions is largely intergovernmental, with political, economic, and environmental exchanges the dominant type of contacts. This level of communication is conducted at three levels. At a bilateral level as demonstrated by Spanish-Moroccan exchanges and Franco-Algerian links, at a multilateral-bilateral level as seen in EU-Moroccan negotiations and NATO-Egyptian dialogue, and at a strictly multilateral level, as demonstrated in EU-AMU negotiations. North-South transnational forces are mainly active in the energy sector with the construction of two pipelines linking the Maghreb to Western Europe.

Other contact points between the three subregions bordering the Mediterranean take place in the military and social areas. The picture that emerges when the lens is cast upon military interaction in the basin is a somewhat balanced one. Southern Europe, the Maghreb and the Levant are all net importers of military hardware. Yet, that is where similarities terminate. The northern shore of the basin is far superior when it comes to the technological sophistication of military equipment. Exchanges are predominantly intergovernmental but transnational interaction, which includes an increase in arms smuggling and non-state terrorism, is likely to increase if no strict measures are introduced to check the proliferation of such activities.

Social patterns of intercourse between the three subregions bordering the Mediterranean are limited in nature. The narrow range of social links is limited to fields such as tourism (Southern Europe to Maghreb main type of interaction); commerce (workers' remittances from Southern Europe to the Maghreb and the Levant); and education (scholars of both the Maghreb and Levant to Europe) (see Ayubi, 1995: 387-90). The main development in this sector during the last decade is that both the Maghreb and Levant increasingly consider the European integration experiment as a negative undertaking. Restrictive immigration legislation and the rise of extreme right wing political groupings across Western Europe have reinforced the notion among some observers that a "fortress Europe" scenario is emerging.

An assessment of the patterns and level of interaction between the actors in the Mediterranean space therefore reveals an interactive system which is highly asymmetrical. The Mediterranean is a frontier separating different political, economic, military and cultural forces. Instead of acting as a catalyst for an increase in subregional co-operative patterns of interaction, the fluid post-Cold War game rules have led to a more transparent picture of Mediterranean divergences. Mediterranean countries realize that they cannot sensibly ignore their Mediterranean dimension, but they are also aware that extra-Mediterranean ties are an important interface between subregional co-operation and conflict. While all countries in the basin have come to terms with this realist perspective, they differ in their foreign policy approaches towards this "new" geostrategic environment.

Mediterraneo-centric schools of thought advocate that Mediterranean solidarity will precede their fundamental loyalties towards Europe or the Arab world or Islam. But reality depicts a picture where Arab countries are Arab before they are Mediterranean; similarly, Southern European countries are European first and then Mediterranean; even Israel seems to be commencing a process of further integration with Western Europe in an economic co-operation zone. Based on this observation, the patterns of relations in the Mediterranean look like a complex construct of geographical and functional layers. The concept of regional arrangements as a framework for collective security and co-operation is becoming more popular among the great powers and is increasingly welcomed at the regional level. Given the heterogeneous nature of the countries in the Mediterranean area (different colonial history, cultural and linguistic background, plus the lack of a common civic culture within several of the countries), it appears realistic to expect that economic as well as geopolitical factors will push this grouping of states towards the EU, either as a group or as single countries. One can draw a parallel here with the systemic changes taking place between the Caribbean and Central America vis-à-vis the North American Free Trade Area (NAFTA) which embraces the United States, Canada, and Mexico (Hettne, 1994: 146).

Such strong extra-Mediterranean solidarities suggest that attempts to establish the structures, or the foundations for regional co-operation in the Mediterranean are bound to fail. The countries in the littoral have little experience of multilateral institutions. The Western Mediterranean Forum remains a limited political institution that has so far failed to instil any notion of a Mediterranean identity. Reality reveals that Mediterranean issues are being addressed in already existing European and Arab security and co-operation institutions, such as the European Union, the OSCE, and the Arab League. While sound in theory, the plethora of pan-Mediterranean initiatives put forward to date have met with very little success in practice.

International regions have ontological status: they reflect a real patterning of regional interaction, and are more than just an analytical tool. One can argue about the correct interpretation of the dividing lines, but one cannot apply the term "international region" to any group of states. Thus, while one can speak of a European international region that is transnational and intergovernmental dominant, and an intergovernmental Middle Eastern region, there is no evidence of a Mediterranean international region of any type. As things stand, the Mediterranean has become a boundary between the Western European and Middle Eastern regions, whereas for much of classical history it was the centre of an international region.

Theoretically, three possible scenarios of regional development can be mentioned: "regional restabilization" under an external hegemon, in this case either the United States or the European Union; regional fragmentation, peripheralization, and bilateralization of internal and external relations which would be intergovernmental dominant; and a neo-regional alternative implying regional restructuring based on symmetrical and solidarity-oriented patterns of development which would consist of both intergovernmental and transnational patterns of relations (Martin, 1991: 115-34). When one considers the diverse interregional relations and the different internal patterns of interaction between and within the subregions encompassing the Mediterranean, and the fact that this area has been increasingly marginalized in the structure of the world system as a whole, regional co-operation with a hegemonic actor which possesses a strong economic base, such as the United States or the EU, seems the only realistic positive way forward for the countries in the area. Without external support, states in the Mediterranean could become "front states" between the Western European and Middle Eastern international regions.

To date, the enclosed sea of the Mediterranean is simply a substitute for land frontiers between states. Similar to a land frontier which constitutes a point of demarcation between two states, the Mediterranean sea, an international waterway, remains a divide between three subregional zones, Southern Europe, the Maghreb and the Levant. It is currently a boundary zone of mutual indifference and lack of interaction which could become either a conflict or co-operative defined region depending on which, if any, patterns of interaction intensify across the area. Contemporary references to the notion of a Mediterranean international region of whatever type is nothing more than an illusion. In fact, this point touches upon the fundamental irony that in the Mediterranean there is no country than can actually be labelled a strictly Mediterranean state. Even the micro-state of Malta regards itself as a European country and aspires to full EU membership.

Despite the fact that the littoral is the great cross-roads of history, religion, culture and civilizations, there is no Mediterranean identity. Mediterranean littoral states continue to share responsibility over their common frontiers, but this does not replace their other commitments and interests which arise out of their history and culture, and which are conditioned by their political and economic choices.

The absence of a Mediterranean identity leaves the *mare nostrum* vulnerable to intrusive overlay, like that which dominated the area throughout the Cold War. The candidate most likely to fill the vacuum left by the disappearance of the Soviet Union and the limited withdrawal of the Americans, is the European Union. Through its multiplicity of initiatives in the area, particularly those launched at the Euro-Mediterranean Conference in November 1995, the European Union has succeeded in establishing a dynamic pattern of interaction with the majority of non-member states across the basin. Like the superpowers during the Cold War, the hegemonic presence of the EU in this area reduces the possibility of a comprehensive international region, covering both intergovernmental and transnational spheres, from emerging (Vayrynen, 1984: 342; Kaiser, 1968: 38-51). The participation of a hegemonic actor in the Mediterranean also increases the probability of aggravating existing intra-regional tensions and polarization (Acharya, 1992: 15). In time, the already booming South-East Asian economies will soon be joined by the rapidly expanding Chinese and Indian markets in the search for external sources of energy to sustain their phenomenal levels of growth. Thus, they could also join in the geostrategic competition for sources of energy in the area (see Hitti, 1994: 88).

In summary, the post-Cold War international system is somewhat more fragmented than it was during the Cold War now that superpower overlay has disappeared. This is clearly illustrated by Mediterranean subregional patterns of interaction. During the Cold War, riparian countries either pledged allegiance to one of the superpowers or else opted to remain equidistant between both superpowers by joining the non-aligned movement. Geo-political parameters were firmly established, enabling countries to formulate a long-term foreign policy agenda.

The Mediterranean remains an important international waterway. Intrusive action in the basin is likely to increase again at the start of the twenty-first century, with the Europeans replacing the Americans and Soviets as primary actors. The presence of European multilateral organizations makes it highly unlikely that an autonomous Mediterranean international region will be able to evolve. More realistic alternative scenarios include:

i) if co-operative patterns of subregional and international relations develop, the basin appears set to become an integral part of a European international region. In order to avoid economic and political fragmentation, countries in the Mediterranean may attempt to sort out their differences within the already established EU regional framework. The overarching image is of one centre, Brussels, and concentric circles, one of which would extend across the Mediterranean (see Waever, 1995: 7). The level of "regionness" would be low, resembling the "variable geometry" design of integration. If such an outcome were to develop, the EU could consolidate its sphere of influence as far east as Serbia and as far south as the sub-Sahara;

ii) if conflictual patterns of subregional and international relations evolve, then the Mediterranean is more likely to become a hinterland between competing centres of power. If the Pax Americana gradually gives way to a tricontinental order consisting of NAFTA, the EU and the Asian Pacific region, the Mediterranean area could become one of the frontiers where tensions and conflicts, if not wars, will develop. As an international transit zone to the Gulf, the Mediterranean could become a flashpoint of tension as the three international centres of power seek to maintain access to these critical sea-lines of communication. The international resurgence of religious fundamentalism, particularly that of Islam, is another factor that could lead to a new "cold war" between the countries in the Mediterranean. Certain scholars have predicted that a clash of civilizations is the next ideological confrontation the world will have to contend with (Huntington, 1993). If such a scenario emerges, the Mediterranean could again find itself a frontier zone between a Christian North and Islamic South.

All post-Cold War forces point towards fragmentation in the Mediterranean. While international region designs continue to be discussed and examined, subregional dynamics such as those involving Greece-Turkey-Cyprus, Morocco-Spain, Algeria-France, and Libya-Western Europe, indicate the dominance of subregional patterns of relations. Thus, there is no sign to denote the emergence of any type of international region across the Mediterranean. International regions exist around, but not across the basin. Rather than being credited for attempting to nurture pan-Mediterranean patterns of interaction, it seems more accurate to regard trans-Mediterranean security initiatives as efforts to manage the Mediterranean frontier that separates the existing European and Middle Eastern international regions.

This study emphasizes the significance of international regions as an intermediate level of analysis between the nation-state and the global international system. It posits the existence of international regions as units of analysis which enable a greater understanding of recent political and social change. An important initial task was to define an "international region". In an attempt to include the necessary and sufficient conditions for labelling a collection of states an international region, the following guidelines were adopted: physical contiguity is a necessary characteristic; the particular nature or intensity of interactions among a group of states is what distinguishes them from other groupings; the influence of external states is considered as this is what makes them "international". The definition presented in this research was specifically constructed to avoid a situation in which the term "region" becomes so inclusive that it became useless. When applied to the Mediterranean area, the scheme set out showed that two distinct international regions exist around the basin, Western Europe and the Middle East.

The next fundamental task was to offer a precise description of regional tranformation. In order to understand the internal and external attributes of regional development, this study offered an analytical framework designed to highlight the scope of interaction within and between international regional systems. The model proposed focused on the main types of interaction that can take place, that is, transnational, intergovernmental, and comprehensive pattern of relations. The degree to which a group of states resembles one ideal type of international region more than another was therefore determined by examining the patterns of interaction taking place. Analysis of the three pattern variables assisted in the task of regional boundary delineation as the edge of one region and the start of the next one was identified according to increases or decreases in the intimacy of interaction among the participating states. It also served as a checklist function in the empirical analysis to help detect any fluctuations in the intensity of relations across the Mediterranean area. The level of power of states in the Mediterranean was established which allowed for analysis of all conditions of amity and enmity.

A number of new elements and angles on the regional analysis problem have emerged in this study. The first concerns the unit of analysis itself. An historical review of patterns of interaction in the Mediterranean shows that both co-operative and conflict based regional designs existed and that both types can be as durable as the other. Another structural finding, which does not fit easily into the analytical model devised, is that which manifested itself during the initial phase of the Christian-Muslim period when the

Mediterranean acted as a buffer zone between North and South. Somewhat mirroring contemporary relations in the Mediterranean, the lack of extensive contact across the area relegates this space from international region classification. In the Christian-Muslim period, the eventual intensification of antagonistic relations generated a conflict based region. In more recent times, no such intensive co-operative or conflictual patterns of relations have yet emerged.

The historical review also showed that an increase in intrusive action throughout the Mediterranean, first by such actors as the British, and later by the two superpowers, relegated the Mediterranean area out of the 'international region' taxonomy altogether. Despite an increase of regional rhetoric since the end of the Cold War, the basin remains a zone of relative mutual indifference which lacks intensive indigenous patterns of interaction. Moreover, bipolar overlay has gradually been superseded by a more multipolar intrusive system which manifests itself in the form of international organizations. In the case of the Mediterranean, the United States and the former Soviet Union have already been superseded by the European Union in economic affairs. In addition, the OSCE and NATO are also playing more active political and military roles respectively. Multilateral relations are strong around, but not across the Mediterranean. A review of international organizations operating in the area reveals an overlapping pattern of institutions in the north, namely the EU/WEU, NATO and the OSCE, a more limited number of such groupings in the south, namely the Arab League and the AMU, and no trans-Mediterranean institutional framework which could help nurture pan-Mediterranean relations.

As stated in the proposed definition of an international region, this study has focused on what influence external actors can have on regional transformation processes in order to provide a panorama of the relations at work. This analysis showed that the support or withdrawal of support of an external patron can alter the balance of power in an international region. Intrusive powers are often as important in limiting conflict and even, on occasion, encouraging co-operation, as they are in intensifying strife. The continuous diffusion of power to the peripheral members of the international system has therefore further reduced the dominance formerly exercised by the major powers. The post-Cold War rise of regionalism can thus be viewed as another step in the process of global and regional transformation. Hegemonistic and subimperialistic relationships between great powers and secondary states have gradually been replaced by a dualistic relationship in which major powers are still at the top of the international hierarchy, but are no longer able to steer the development of international economic and political development. Such developments are now also being shaped by semiautonomous regional and domestic forces. The lack of data available in this area makes it apparent that extensive research on the role multilateral actors play in international regions is necessary, along with a re-assessment of superpower patron-client relations in the Mediterranean so that a more complete picture of intrusive behaviour is attained.

This study has closed some of the gaps in the secondary literature on regions. It attempts to follow up on the Cantori and Spiegal comparative framework for the study of regional international relations (Cantori and Spiegal, 1970). Although the international region and subregional unit of analysis cannot always explain international events, it is certain that the inclusion of this level of analysis helps arrive at a more coherent understanding of international relations. There are several advantages to adopting this approach. Scholars are able to analyze events at an intermediate level instead of having to assess relations at an international or state level. The concept of regionalism permits area specialists to apply their expertise in a broader theoretical manner. As this study has shown, the concept of regionalism also permits scholars to conduct a comparative study of both contemporary and historical international regions. Chapter three was dedicated to mapping out regional developments in the Mediterranean through the ages so that a comparative analysis could be conducted with contemporary relations. The notion of international regions also enables scholars to study the interactions of various levels of the international system. For example, in addition to examining the dynamics of specific international regions, regionalists are also able to assess how international regions interact with each other.

The fact remains that the importance of the regional level of analysis in the discipline of International Relations continues to be underplayed. This study attempts to offset this tendency. International regions go against the grain of the Anglo-American analytical tradition in International Relations which developed an insular perspective of international affairs by concentrating on the interaction between reigning great powers. The diffusion of power in the contemporary system raises the importance of the states at the lower levels of the power hierarchy. Understanding how these actors interact with one another and with other actors higher up the power ladder is essential if the dynamics of the multipolar international system are to be better understood.

This study has provided an holistic approach to correct the imbalance in the secondary literature which has been dominated by coverage of the Mediterranean from either a domestic or international vantage point. The framework of analysis has struck a balance between these two approaches which enabled a comprehensive assessment of international regional development. However, when applied to the case study of the Mediterranean, this holistic approach to international relations (political, economic, social, environmental and military), proved problematic and provided inconclusive scientific results. One of the reasons for this was the unavailability of standardized statistics on interaction between the countries in the Mediterranean. This hindrance was compounded by the more general fact that post-Cold War regional dynamics remain fluid and it is therefore difficult to test the results of such investigations into regional transformation. In retrospect, a more incremental approach to the concept of regionalism would have probably been a more productive basis to start assessing such diverse units of international relations as those which exist in the Mediterranean area. Scholars of international relations should limit their research to specific areas of regional development, for example, focusing on indigenous patterns of interaction, patterns of relations with neighbouring regions, and the role of non-regional actors in an international region. Only after such specialized investigations have been systematically carried out should attempts at providing a more complete panorama of the full network of relations be undertaken. This will help avoid achieving complicated and vague results which led to the initial abandonment of the concept of regionalism.

The analysis of intergovernmental and transnational patterns of relations worked well until it came to explaining what impact such forces have on one another. Research into transnational activities has not kept up with the increase this type of interaction has registered in recent years. The absence of such information makes it difficult, if not impossible, to conduct a comparative analysis of transnational and intergovernmental patterns of regional relations.

In addition, this research exercise highlighted several areas ripe for political and scholarly attention. For example, how durable are the international regions in question, and what is the relationship between their internal dynamic, and their interaction with other international regions? How do the exchanges within the structure of international regions influence the foreign policy options available to the states concerned? And what, if any, real leverage do the states concerned have over the course of events in their region? Application of the theoretical framework also clearly showed that further analysis is required in the areas of boundary delineation, subregional configurations, and the definition of regionalism. No rigorous blueprint has yet been developed and tested which would facilitate the difficulty in drawing boundaries between regions. Similarly, more insight into subregional configurations is required so that a better understanding of subregional patterns of development can be achieved. The definitional model used for the concept of regionalism in this analysis must be regarded as a first step in the right direction. More attention needs to be dedicated to identifying whether the inclusion of other attributes in the definition of regionalism would shed a more accurate picture of the regional transformation process.

At the end of the twentieth century the world is perhaps best described as a dynamic multipolar international system based on international regions and subregions within which the struggles of nationalist identities and hegemonic regional leadership rivalries continue to occur (Cantori, 1994: 22). The collapse of superpower influence in international affairs and the multiplication of independent states has helped the notion of region to become a crucial unit of analysis in the post-Cold War era. The most recent phase of regionalism is a manifestation of the effective spread of power from the core of the international system to its peripheries. It has spurred the emergence of new regional power centres, integration schemes, and regional conflict formations. All patterns of relations are present in the Mediterranean area. It therefore seems logical that regional theory should be revived as a way of looking at the post-Cold War world.

This study categorizes and illustrates patterns and processes at work in the intermediate arena of the international system - the international region. This research project also goes some way toward mapping out a path that other regionalists can follow to help comprehend more clearly the concept of regionalism. A regionalist should consider which states are likely to integrate or disintegrate, which states could become allies or adversaries, what the content of future disputes might be, which states are likely to emerge as regional power centres, and what role outside powers are likely to play. By the nature of these concerns, a region systems theorist considers what the effects of these developments are likely to be on the regional, as well as the global, balance of power. All this research and theorizing on international regional development and the place of this unit within the world political system makes one implication clear. Knowledge to cope with the concept of regionalism is still lacking. The analytical framework in this study provides a foundation for a better understanding of regional transformation in the post-Cold War period. Developing a more rigorous theoretical framework will have to wait until the complexities and overlaps and data problems encountered are overcome.

The schema provided therefore strikes a balance between the two fundamental levels of regional analysis, the internal and external dynamic approaches of regional systems. The three ideal types of international regions equips the student of regions with the necessary tools to categorize and illustrate the complex features at work at a regional level. Although it is necessary to further elucidate the analysis presented before a synthesized and standardized definition of the international region and more conclusive propositions concerning regional transformation can be achieved, the scheme of analysis put forward demonstrates that it is possible to construct an operational comparative framework for the study of international regions. It is hoped that this study will attract a wider arena of research and reflection on the development of international regions so that further progress in this direction is achieved.

Appendix

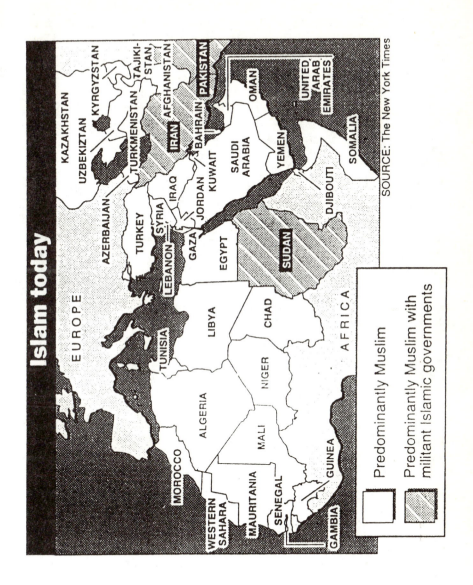

Islam today

SOURCE: The New York Times

Predominantly Muslim

Predominantly Muslim with militant Islamic governments

245

Bibliography

Reference Books

British Yearbook of International Law (1988), Ago, R., "The First International Communities in the Mediterranean World".

Dictionary of the Social Sciences (1964), Self, P. J., "Region" and "Regionalism", Julius Gould and William L. Kolb (eds.), Tavistock Publications.

Encyclopaedia of the Social Sciences (1953), Hintze, H., "Regionalism" in Elwin, R. A. Seligman and Alvin Johnson (eds.), Vol. 13, Macmillan Company.

International Encyclopaedia of Social Sciences (1968), Isard, W. and Thomas A. Reiner, "Region", David L. Sills (ed.), Macmillan Company and Free Press, Vol. 13.

International Encyclopaedia of Social Sciences (1968), Vance, R. B. "Region", Sills, D. (ed.), Macmillan Group and the Free Press, Vol. 13.

International Encyclopaedia of Social Sciences (1968), Kaplan, M., "International Systems", David L. Sills (ed.), New York: The Macmillan Company and Free Press, Vol. 15.

The Middle East and North Africa (1993), Joffe, G., "The Implications of the 'New World Order' for the Middle East and North Africa", Europa.

Times Atlas of World History (1984), Barraclough, G. (ed.), Times Books.

Books

Acharya, A. (August 1993), "A New Regional Order in South-East Asia: ASEAN in the Post-Cold War Era", Adelphi Paper No. 279.

Agarwal et al. (March 1994), "EC Economic Integration and its Impact on Foreign Direct Investment and Developing Countries", in Ohno, K. and Okamoto, Y. (eds.), *Regional Integration and Foreign Direct Investment: Implications For Developing Countries*, Tokyo, Japan.

Agha, H. (1994), "The Middle East and Europe: the Post-Cold War Climate", in Miall, H. (ed.), *Redefining Europe,* Pinter Publishers.

Ahmed, A. S. (1992), "Les relations energetiques CEE-Maghreb" in Dumas, M. (ed.) (1992), *Mediterranee Occidentale: Securite et Cooperation*, Paris: Foundation pour les etudes de defense nationale.

Ahmed, A.S. (1995), "Economic and energy relations between Europe and the Maghreb countries", in Ayubi, N.N. (ed.) (1995), *Distant Neighbours*, Ithaca Press, Reading.

Ajami, F. (1981), *The Arab Predicament*, Cambridge University Press, Cambridge.

Aliboni, R. (March 1991), "European Security Across the Mediterranean", *Challiot Papers*, No. 2, Institute for Security Studies of Western European Union.

Aliboni, R. (ed.) (1992), *Southern European Security*, Pinter Publishers.

Anderson, K. and Blackhurst, R. (eds.) (1993), *Regional Integration and the Global Trading System*, Harvester Wheatsheaf.

Anderson, K. and Norheim, H. (1993), "History, Geography and Economic Integration", in Anderson, K. and Blackhurst, R. (eds.) (1993), *Regional Integration and the Global Trading System*, Harvester Wheatsheaf.

Awad, I. (1994), "The Future of Regional and Subregional Organization in the Arab World", in Tschirigi, D. (ed.), *The Arab World Today*, Lynne Rienner

Publishers, London.

Ayubi, N.N. (ed.) (1995), *Distant Neighbours, The Political Economy of Relations between Europe and the Middle East/North Africa*, Ithaca Press, Reading.

Blacksell, M. (1984), "The European Community and the Mediterranean Region: Two Steps Forward, One Step Back", in Williams, A., *Southern Europe Transformed*, Harper and Row Publishers, London.

Blin, L. and Benoit Parisot (1992), "Les Relations economique entre la CEE et les pays du Maghreb", Kacem Basfao and Jean-Robert Henry (eds.), *Le Maghreb, L'Europe et la France*, Paris: CNRS.

Braudel, F. (1973), *The Mediterranean and the Mediterranean World* [translation of "La Mediterraneanee et le Monde Mediterraneanee"], Vol. II, London, Collins.

Brown, L. C. (1984), *International Politics and the Middle East,* Princeton University Press.

Busch, M. L. and Milner, H. V. (1994), "The Future of the International Trading System: International Firms, Regionalism, and Domestic Politics", in Stubbs, R. and Underhill, G. R. D. (eds.) (1994), *Political Economy and the Changing Global Order*, Macmillan.

Busuttil, S., Calleja, J., Wiberg, H. (eds.) (1994), *The Search for Peace in the Mediterranean Region*, Mireva Publications.

Buzan, B. and Rizvi, G., et al. (1986), *South Asian Insecurity and the Great Powers*, London: Macmillan.

Buzan, B. (1991a), *People, States and Fear: An Agenda for International Security Studies in the post-Cold War Era*, Hemel Hempstead, Harvester Wheatsheaf.

Buzan, B. and Roberson, B.A. (1993), "Europe and the Middle East: Drifting Towards Societal Cold War?" in Waever et al., *Identity, Migration and the New Security Agenda in Europe*, London, Pinter Publishers.

Buzan, B., Charles Jones and Richard Little (1993a), *The Logic of Anarchy: Neorealism to Structural Realism*, New York, Columbia University Press.

Buzan, B., Waever, O., Wilde, J., et al. (1995, Draft),"Regional Security:A Post-Cold War Framework for Analysis", forthcoming.

Calleya, S. (1994a), "Malta's Post-Cold War Perspective on Mediterranean Security", in Gillespie, R. (ed.), *Mediterranean Politics*, Vol. 1, Pinter Publishers.

Calleya, S. (1994b), "Conflict Prevention and Peace-Building Measures in the Mediterranean", in Busuttil, S., Calleja, J. and Wiberg H. (eds.), *The Search For Peace in the Mediterranean Region*, Mireva Publications.

Calleya, S. (1994c), "Prospects for EC Expansion in the Mediterranean Region", The SouthEast European Yearbook 1993, Hellenic Foundation for European and Foreign Policy, Athens 1994.

Calleya, S. (1996), "Reassessing Mediterranean Security" in House of Commons Defence Committee Third Report, "Nato's Southern Flank", London, HMSO, 13 March 1996.

Cantori, L. J. and S. L. Spiegal (1970), *The International Politics of Regions: A Comprehensive Approach*, Eaglewood Cliffs, N. J. Prentice-Hall.

Carle, C. (1992), "France, the Mediterranean and Southern European Security", in Aliboni, R. (ed.) (1992), *Southern European Security*, Pinter Publishers.

Claude, I. (1959), *Swords into Ploughshares*, New York: Random House.

Cobb, R. and Charles Elder (1970), *International Community: A Regional and Global Study*, New York: Holt, Reinhart and Winston.

Colbert, E. (1992), "Southeast Asian Regional Politics: Toward a Regional Order", in Wriggins, W. H. (ed.), *Dynamics of Regional Politics*, Columbia University Press, New York.

Cremasco, M. (1984), "The Military Presence of the Riparian Countries", in

Luciani, G. (ed.), *The Mediterranean Region*, Croom Helm Ltd.

Dawisha, A. and Zartman, I. W. (eds.) (1988), *Beyond Coercion. The Durability of the Arab State*, London, New York and Sydney: Croom Helm.

Deutsch, K. W. (1954), *Political Community at the International Level, Problems of Measurement and Definitions*, Garden City, N.Y: Doubleday.

Deutsch, K. W., et al. (1957), *Political Community and the North Atlantic Area*, Princeton.

Deutsch, K. W. (1966), *Nationalism and Social Communication*, Cambridge, Mass.: The MIT Press.

Dumas, M. (ed.) (1992), *Mediterranee Occidentale: Securite et Cooperation*, Paris: Foundation pour les etudes de defense nationale.

Etzioni, A. (1965), *Political Unification, A Comparative Study of Leaders and Forces*, New York: Holt, Reinhart and Winston.

Fleming, W. G. (1969), "Sub-Saharan Africa: case studies of international attitudes and transactions of Ghana and Uganda", in J. N. Rosenau (ed.), *Linkage Politics: Essays on the Convergence of National and International Systems*, New York: Free Press.

Gibb, R. (1994), "Regionalism in the World Economy", in Gibb, R. and Michalak, W. (eds.), *Continental Trading Blocs*, John Wiley & Sons Publishing.

Gibb, R. and Michalak, W. (eds.) (1994), *Continental Trading Blocs, The Growth of Regionalism in the World Economy*, John Wiley & Sons Publishing.

Gillespie, R. (ed.) (1994), *Mediterranean Politics*, Pinter Publishers.

Greco and Guazzone (1992), "Continuity and change in Italy's security policy", in Aliboni, R. (ed.) (1992), *Southern European Security*, Pinter Publishers.

Greenpeace (1992), "Proposals for an Integrated Security in the

Mediterranean", Greenpeace Proyecto Mediterraneo.

Henson, P. (1994), "The European Union and the Mediterranean", Workshop Report, The University of Reading Press.

Hettne, B. (1994), "The regional factor in the formation of a new world order", in Sakamoto, Y. (ed.), *Global Transformation, Challenges to the State System*, UN University Press, Tokyo, 1994.

Hinnebusch, R. A. (1991), "The Foreign Policy of Syria", in Korany, B. and Dessouki, A. E. H. (eds.) (1991), *The Foreign Policies of Arab States*, Westview Press.

Hinsley, F. H. (1963), *Power and the Pursuit of Peace*, Cambridge University Press.

Hitti, N. (1994), "The internationalization of the state in the Middle East", in Sakamoto, Y. (ed.), *Global Transformation, Challenges to the State System*, UN University Press, Tokyo.

Holland, R. F. (1985), *European Decolonization 1918-1981, An Introductory Survey*, Macmillan Distribution Ltd., London.

Holmes, J. W. (1995), "U.S. Interests and Policy Options", in Holmes, J. W. (ed.), *Maelstrom, The United States, Southern Europe, and the Challenges of the Mediterranean*, The World Peace Foundation.

Holsti, K. J. (1977), *International Politics*, Eaglewood, Cliffs, N. J. Prentice-Hall.

Italian Ministry of Foreign Affairs (1991), *The Mediterranean and the Middle East after the War in the Gulf: The CSCM*, Rome.

Joffe, G. (1994), "The European Union and the Maghreb", in Gillespie, R. (ed.), *Mediterranean Politics*, Pinter Publishers.

Kaplan, L. and Clawson, R. W. (1985), "East-West Competition and the Mediterranean Powers in Historical Perspective", in Kaplan, L., Clawson, R. W. and Luraghi, R. (eds.), *NATO and the Mediterranean*, Wilmington,

252   *Navigating Regional Dynamics in the Post-Cold War World*
Scholarly Resources Inc.

Kegley, Jr., C. W. and Raymond, G. (1994), *A Multipolar Peace*, St. Martin's Press.

Kennedy, P. (1988), *The Rise and Fall of Great Powers*, Penguin Publishers.

Kennedy, P. (1993), *Preparing For The Twenty-First Century*, New York: Random House.

Kerdoun, A. (1994), "The Maghreb and the Problem of Security and Cooperation for Development in the Mediterranean", in Busuttil, S., Calleja, J., Wiberg, H. (eds.), *The Search for Peace in the Mediterranean Region*, Mireva Publications.

Khader, B. (1995), "Euro-Arab trade relations", in Ayubi, N.N. (ed.) (1995), *Distant Neighbours*, Ithaca Press, Reading.

Koenigsberger, H. G. (1987), *Medieval Europe 400-1500*, Longman Group UK Limited.

Korany, B. and Dessouki, A. E. H. (eds.) (1991), *The Foreign Policies of Arab States*, Westview Press.

Krasner, S. (ed.) (1983), *International Regimes*, Ithaca, N.Y. and London: Cornell University Press.

Krasner, S. (1985), *Structural Conflict*, Berkeley: University of California Press.

Latter, R. (1992), *Mediterranean Security*, Conference Report based on Wilton Park Conference 372: 21-25 October 1991: "Mediterranean Security, Uncertainties and Opportunities in a Changing World", London, HMSO.

Lesser, I.O. (1992), *Mediterranean Security: New Perspectives and Implications for U.S. Policy*, Santa Monica, California: Rand.

Lindberg, L. and Scheingold, S. (eds.) (1970), *Regional Integration: Theory and Research*, Cambridge, Mass.: Harvard University Press.

Liska, G. (1968), *Alliances and the Third World*, Baltimore: Johns Hopkins Press.

Luciani, G. (ed.) (1984), *The Mediterranean Region*, Croom Helm, London.

Luciani, G. and Salame, G. (eds.) (1988), *The Politics of Arab Integration*, Croom Helm, London.

Luciani, G. (1995), "Euro-Maghreb co-operation and conflict in the areas of oil and gas", in Ayubi, N.N. (ed.) (1995), *Distant Neighbours*, Ithaca Press, Reading.

Luttwak, E. N. (1975), "Mediterranean Policies in Historical Perspective", in Bassioni, M. S. (ed.), *Issues in the Mediterranean*, Chicago, The Chicago Council on Foreign Relations.

Lyons, T. P. (1992), "The Horn of Africa Regional Politics: A Hobbesian World", in Wriggins, W. H. (ed.), *Dynamics of Regional Politics*, Columbia University Press, New York.

MacFarlene, S. N. (1985), "Intervention and Regional Security", Adelphi Paper No. 196, IISS Publications.

Madgwick, P. J., Steeds, D. and Williams, L. J. (1982), *Britain Since 1945*, London.

Martin, G. (1989), "France and Africa", in Aldrich, R. and Connell, J. (eds.), *France in World Politics*, Routledge, London.

Mearsheimer, J., "Disorder Restored", in Allison, G. and Treverton, G. F. (eds.) (1992), *Rethinking America's Security*, New York: Norton and the Council on Foreign Relations.

Melo, J. and Panagariya, A. (1992), *The New Regionalism in Trade Policy*, Washington D.C.: World Bank Centre for Economic Policy Research.

Miall, H. (ed.) (1994), *Redefining Europe, New Patterns of Conflict and Cooperation*, Pinter Publishers, London.

Michalak, W. (1994), "The Political Economy of Trading Blocs", in Gibb, R. and Michalak, W. (eds.), *Continental Trading Blocs, The Growth of Regionalism in the World Economy*, John Wiley & Sons Publishing.

Mikdashi, Z. (1995), "Issues in oil revenue investments with special reference to Euro-Arab Relations", in Ayubi, N.N. (ed.) (1995), *Distant Neighbours*, Ithaca Press, Reading.

Military Balance 1994-95, IISS, Brasseys.

Miller, L. H. (1970), "Regional Organizations and Subordinate Systems", in L. J. Cantori and S. L. Spiegal (eds.), *The International Politics of Regions: A Comparative Approach*, Englewood Cliffs, N. J.: Prentice-Hall.

Monroe, E. (1938), *The Mediterranean in Politics*, Oxford University Press.

Mortimer, E. (1992), "European Security After the Cold War", Adelphi Paper No. 271, IISS Publications.

Noble, P. C. (1991), "The Arab System: Pressures, Constraints, and Opportunities", in Korany, B. and Dessouki, A. E. H. (eds.) (1991), *The Foreign Policies of Arab States*, Westview Press.

Obdeijn, H. (1995 working draft), "Western Europe's Near Abroad: North Africa", in Waever.

Ohno, K. and Okamoto, Y. (eds.) (1994), *Regional Integration and Foreign Direct Investment: Implications For Developing Countries*, Institute of Developing Economies, Tokyo, Japan.

Ortega, A. (1995), "Relations With The Maghreb", in Holmes, J. M. (ed.), *Maelstrom*, The World Peace Foundation.

Pirenne, H. (1936), *A History of Europe from the Invasions to the XVI Century*, George Allen & Unwin Ltd.

Pomfret, R. (1986), *Mediterranean Policy of the EC, A Study of Discrimination in Trade*, Macmillan, London.

Pomfret, R. (1989), *The European Community: Three Issues*, European Documentation Centre, Malta.

Popper, K. (1964), *The Poverty of Historicism*, New York.

Pryor, J. H. (1988), *Geography, Technology, and War Studies in the Maritime History of the Mediterranean*, 649-1571, Cambridge University Press.

Redmond, J. (1993), *The Next Mediterranean Enlargement*, Dartmouth, Aldershot.

Regelsberger, E. and Wolfgang, W. (1984), "European Concerts for the Mediterranean Region", in Luciani, G. (ed.), *The Mediterranean Region*, Croom Helm Ltd.

Rodrigo, F. (1992), "The end of the reluctant partner: Spain and Western security in the 1990s", in Aliboni, R. (ed.), *Southern European Security*, Pinter Publishers.

Rose, J. H. (1933), *The Mediterranean in the Ancient World*, Ares Publishers Inc., Chicago.

Rosenau, J. N. (ed.) (1969), *International Politics and Foreign Policy: A Reader in Research and Theory*, New York: Free Press.

Rosenau, J. N. (ed.) (1969), *Linkage Politics: Essays on the Convergence of National and International Systems*, New York: Free Press.

Rosenau, J. N. (1990), *Turbulence in World Politics: A Theory of Chance and Continuity*, Princeton: Princeton University Press.

Russett, B. M. (1967), *International Regions and the International System: A Study in Political Ecology*, Chicago, Rand McNally.

Sakamoto, Y. (ed.) (1994), *Global Transformation, Challenges to the State System*, UN University Press, Tokyo.

Salame, G. (1988), "Integration in the Arab World: The Institutional Framework", in Luciani, G. and Salame, G. (eds.) (1988), *The Politics of*

Arab Integration, Croom Helm Ltd.

Santos, H. (1992), "The Portugese national security policy", in Aliboni, R. (ed.) (1992), *Southern European Security*, Pinter Publishers.

Schmitter, P. (1970), "A Revised Theory of Regional Integration", in Lindberg, L. and Scheingold, S. (eds.), *Regional Integration: Theory and Research*, Cambridge, Mass: Harvard University Press.

Sezer, D. B. (1992), "Prospects for Southern security: a Turkish perspective", in Aliboni, R. (ed.) (1992), *Southern European Security*, Pinter Publishers.

Siegfried, A. (1948), *The Mediterranean*, Alden Press.

Sorokin, P. A. (1947), *Society, Culture, and Personality: Their Structure and Dynamics*, N.Y. Cooper Square.

Spencer, C. (1993), "The Maghreb in the 1990s", Adelphi Paper No. 274, Brasseys for IISS.

Spiegal, S. (1985), *The Other Arab-Israeli Conflict: Making America's Middle East Policy From Truman to Reagan*, Chicago, The University of Chicago Press.

Strange, S. (1995), "European direct investment in North Africa: The investor's perspective" in Ayubi, N.N. (ed.) (1995), *Distant Neighbours*, Ithaca Press, Reading.

Stubbs, R. and Underhill, G. R. D. (1994), "Global Trends and Regional Patterns", in Stubbs, R., and Underhill, G. R. D. (eds.) (1994), *Political Economy and the Changing Global Order*, Macmillan.

Stubbs, R. and Underhill, G. R. D. (eds.) (1994), *Political Economy and the Changing Global Order*, Macmillan.

Sundelius, B. (1982), "The Nordic Model of Neighbourly Cooperation", in *Foreign Policies of Northern Europe*, Boulder Press.

Toynbee, A. J. and Jane Caplan (1972), *A Study of History*, New York, N.Y.:

Oxford University Press.

Toynbee, A. J. (1950), *War and Civilization*, New York, N.Y.: Oxford University Press.

Tschirgi, D. (ed.) (1994), *The Arab World Today*, Lynne Rienner Publishers, London.

Vasconcelos, A. (1992), "The shaping of a subregional identity", in Aliboni, R. (ed.) (1992), *Southern European Security*, Pinter Publishers.

Vatikiotis, P. J. (1969), *The Modern History of Egypt*, Weidenfeld and Nicolson Publishers.

Vatikiotis, P. J. (1984), *Arab and Regional Politics in the Middle East*, Croom Helm, London.

Wallace, W. (1990), *The Dynamics of European Integration*, London: Royal Institute of International Affairs/Pinter Publishers.

Waltz, K. N. (1979), *Theory of International Politics*, Addison-Wesley.

Williams, A. (1984), *Southern Europe Transformed, Political and Economic Change in Greece, Italy, Portugal and Spain*, Harper and Row Publishers, London.

Wriggins, W. H. et al. (1992), *The Dynamics of Regional Politics: Four Systems on the Indian Ocean Rim*, Columbia University Press, New York.

Wright, Q. (1965), *A Study of War*, Chicago, Ill., University of Chicago Press.

Zartman, I. W. (1984), "Maghrebi Politics and Mediterranean Conflict", in Luciani, G. (ed.), *The Mediterranean Region*, Croom Helm, London.

Articles

Acharya, A. (1992), "Regional Military-Security Cooperation in the Third World: A Conceptual Analysis of the Relevance and Limitations of ASEAN", *Journal of Peace Research*, Vol. 29, No. 1.

Agence Europe, 9 April 1993.

Agence Europe, 26 June 1994.

Ajami, F. (1978), "The End of Pan-Arabism", *Foreign Affairs*, Winter, 1978-9.

Allen, D. (1978), "The Euro-Arab Dialogue", *Journal of Common Market Studies*, 16 July 1978.

Anyaoku, E. (1994), "The Commonwealth and the New Multilateralism", *The Round Table*, 331, July 1994.

Ash, T. (1995), "Jordan takes Centre Stage", *MEED*, Vol. 39, No. 45, 10 November 1995.

Atherson, Jr., A. L. (1984), "Arabs, Israelis - and Americans: A Reconsideration", *Foreign Affairs*, 62, 5, Summer 1984.

Atlantic News (1993), "WEU: Italy Proposes A Land Component For The Future Southern Air-Naval Force", No. 2574, 26 November 1993.

Ayoob, M. (1979), "The Superpowers and Regional 'Stability': Parallel Responses to the Gulf and the Horn", *The World Today*, Vol. 35.

Balassa, B. and Bauwens, L. (1988), "The determinants of intra-European trade in manufactured goods", *European Economic Review*, 32.

Banks, M. (1969), "Systems Analysis and the Study of Regions", *International Studies Quarterly*, Vol. 13, December 1969.

Bauer, G. E. (1994), "The Morocco-Polisario Conflict: Prospects for Western Saharan Stability in the 1990s", *Small Wars and Insurgencies*, Vol. 5, No. 1,

Spring 1994.

Biad, A. (1993), "The Maghreb's Response to the Gulf War", *Mediterranean Social Sciences Review*, Vol. 1, No. 1, April 1993.

Binder, L. (1958), "The Middle East as a Subordinate International System", *World Politics*, 10, 3, April 1958.

Blair, E. (1995), "Winning Over the World's Investors", *MEED*, Vol. 39, No. 40, 6 October 1995.

Blunden, M. (1994), "Insecurity on Europe's Southern Flank", *Survival*, Vol. 36, No. 2, Summer 1994.

Brecher, M. (1963), "International Relations and Asian Studies: The Subordinate State System of Southern Asia", *World Politics*, 15, 2 January 1963.

Brecher, M. (1969), "The Middle East Subordinate System and Its Impact on Israel's Foreign Policy", *International Studies Quarterly*, 13, 2, June 1969.

Bryant, R. C. (1994), "Global Change", *The Brookings Review*, Fall 1994.

Buzan, B. (1991b), "New Patterns of Global Security in the Twenty-First Century", *International Affairs*, Vol. 67, No. 3.

Buzan, B. and Segal, G. (1994a), "Rethinking East Asian Security", *Survival*, Vol. 36, No. 2, Summer 1994a.

Buzan, B. (1994b), "International Society and International Security", *Millennium*, 7, 1994b.

Buzan et al. (1994c), "Regional Security: A Post-Cold War Framework For Analysis", Working Paper, Centre for Peace and Conflict Research, Copenhagen, 1994.

Buzan, B. and Little, R. (1994d), "The Idea of International System: Theory Meets History", *International Political Science Review*, 15:3.

Calleya, S. (1993), "Europe After Maastricht", in *Sunday Times* (Malta), 26 September 1993.

Calleya, S. (1994d), "Reassessing Mediterranean Security", in *Sunday Times* (Malta), 4 December 1994.

Campbell, J. C. (1980), "Les Etats-Unis et l'Europe au Moyen Orient: interets communs et politiques divergentes", *Politique Internationale*, 7.

Cantori, L. J. and Spiegal, S.L. (1969), "International Regions: A Comparative Approach to Five Subordinate Systems", *International Studies Quarterly*, Vol. 13, December 1969.

Cantori, L. J. and Spiegal, S. L. (1973), "The Analysis of Regional International Politics: The Integration Versus The Empirical Systems Approach", *International Organization*, 27.

Cantori, L. J. (1994), "Regional Solutions To Regional Security Problems: The Middle East And Somalia", *Middle East Policy*, Volume III.

Claes, W. (1995), "NATO and the Evolving Euro-Atlantic security architecture", *NATO Review*, Vol. 42, No. 1, Dec.-Jan. 1995.

Couloumbis, T. A. and Thanos Veremis (1994), "In Search of New Barbarians: Samuel P. Huntington and the Clash of Civilisations", *Mediterranean Quarterly*, Vol. 5, No. 1, Winter 1994.

Cremasco, M. (1979), "NATO's Southern Flank in the East-West Balance", *Lo Spettatore Internazionale*, XIX, 1, Jan.-Mar. 1979.

De Marco, G. (1992), "De Marco calls for the setting up of a Council of the Mediterranean", *Sunday Times* (Malta), 22 November 1992.

De Marco, G. (1993), "Mediterranean Quel Avenir?", *Malta Review of Foreign Affairs*, Special Issue for the CSCE Mediterranean Seminar, May 1993.

Dominguez, J. E. (1971), "Mice That Do Not Roar: Some Aspects of International Politics in the World's Peripheries", *International Organization*, Vol. 25, Spring 1971.

Falk, R. (1995), "Regionalism and World Order After the Cold War", *Australian Journal of International Affairs*, Vol. 49, No. 1, May 1995.

Farley, J. (1994), "The Mediterranean - southern threats to northern shores?", *The World Today*, Vol. 50, No. 2, February 1994.

Fenech, D. (1993), "East-West to North-South in the Mediterranean", *Geo Journal*, 31.2, Kluwer Academic Publishers, October 1993.

George, B. (1993), "The Alliance at the Flashpoint of a New Era", *NATO Review*, No. 5.

Ghebali, V. Y. (1993), "Toward a Mediterranean Helsinki-Type Process", *Mediterranean Quarterly*, Vol. 4, No. 1, Winter 1993.

Ghiles, F. (1992), "The Arab Maghreb Union: Impending Demise?", *Middle East International*, No. 9, 9 October 1992.

Grant, R. J., Papadakis, M. C. and Richardson, J. D. (1993), "Global Trade Flows: Old Structures, New Issues, Empirical Evidence", *Journal of North American Issues*.

Haas, E. B. (1958), "The Challenge of Regionalism", *International Organization*, Vol. 12, No. 4.

Haas, E. B. (1970), "The Study of Regional Integration: Reflections on the Joy and Anguish of Pretheorizing", *International Organization*, Vol. 24, Autumn 1970.

Haas, M. (1970), "International Subsystems: Stability and Polarity", *American Political Science Review*, Vol. 64, March 1970.

Helleiner, E. (1994), "Regionalization in the International Political Economy: A Comparative Perspective", *Eastern Asia Policy Papers*, No. 3, September 1994.

Hellmann, D. C. (1969), "The Emergence of an East Asian International Subsystem", *International Studies Quarterly*, Vol. 13, December 1969.

Hockenos, P. (1994), "Arms Bizarre", *In These Times*, Vol. 18, No. 19, 8 August 1994.

Huntington, S. P. (1993), "The Clash of Civilizations?", *Foreign Affairs*, 72, Summer 1993.

Huntington, S. P. (1993), "If Not Civilizations, What? Paradigms of the Post-Cold War World", *Foreign Affairs*, Vol. 72, No. 5, Nov./Dec. 1993.

Hurrell, A. (1992), "Latin America in the New World Order: A Regional Bloc of the Americas?" *International Affairs*, Vol. 68, No. 1, January 1992.

Ispahani, M. Z. (1983-84), "Alone Together: Regional Security Arrangements in South Africa and the Arabian Gulf", *International Security*, Vol. 8.

Kaiser, K. (1971), "Toward a Theory of Multinational Politics", *International Organization*, No. 4, Autumn 1971.

Kaiser, K. (1968), "The Interaction of Regional Subsystems: Some Preliminary Notes on Recurrent Patterns and the Role of Superpowers", *World Politics*, Vol. 21, 1, October 1968.

Kemp, W. (1994), "Giving Teeth to the CSCE?", *The World Today*, Vol. 50, No. 10, October 1994.

Kennedy, P. (1993), "Preparing For the 21st Century: Winners and Losers", *New York Review of Books*, XL, No. 4, 11 February 1993.

Khader, B. (1991), "Immigration maghrebine face a l'Europe 1992", *Migrations Societe*, Vol. 3, No. 15.

Lambert, J. (1971), "The Cheshire Cat and the Pond: EEC and the Mediterranean Area", *Journal of Common Market Studies*, Vol. X.

Lok, J. J. (1991),"Carrier cuts will hit Sixth Fleet", *Jane's Defence Weekly*, Vol. 16, No. 17, 26 October 1991.

Mace, G., L. Belanger and J. P. Therien (1992), "Regionalism in the Americas and the Hierarchy of Power", *Journal of InterAmerican Studies and World*

Affairs.

Marchand, M. H. (1994), "Gender and New Regionalism in Latin America: Inclusion/Exclusion", *Third World Quarterly*, Vol. 15, No. 1.

Martin, W. G. (1991), "The Future of Southern Africa: What Prospects After Majority Rule", *Review of African Political Economy*, No. 50.

Michalet, C. A. (1994), "Transnational Corporations and the Changing International Economic System", *Transnational Corporations*, Vol. 3, No. 1, UN Publications, February 1994.

Modelski, G. (1961), "International Relations and Area Studies: the Case of South-East Asia", *International Relations*, 2, April 1961.

Mortimer, E. (1991), "Christianity and Islam", *International Affairs*, Vol. 67, No. 1.

Neumann, I. B. (1994), "A Regional-Building Approach to Northern Europe", *Review of International Studies*, Vol. 20, No. 1, January 1994.

Nye, J. S. Jr. (1992), "What New World Order?", *Foreign Affairs*, Vol. 71, No. 2, Spring 1992.

Nye, J. (1970), "Comparing Common Markets: A Revised Neo-Functionalist Model", *International Organization*, Vol. 24, No. 4, Autumn 1970.

Nye, J. (1965), "Patterns and Catalysts in Regional Organization", *International Organization*, Vol. 19, No. 4, Autumn 1965.

Ohmae, K. (1993), "The Rise of the Region State", *Foreign Affairs*, Spring 1993.

Ordonez, F. (1990), "The Mediterranean: Devising a Security Structure", *NATO Review*, 38, October 1990.

Pace, R. (1993), "The New Mediterranean Policy of the European Community", *Bank of Valletta Review*, No. 7, Spring 1993.

Puchala, D. (1968), "The Pattern of Contemporary Regional Integration", *International Studies Quarterly*, Vol. 12, March 1968.

 Rato, R. (1994), "Co-operation and Security in the Mediterranean", North Atlantic Assembly, AL223 PC/SR (94) 2.

Redmond, J. (1993), "The European Community and the Mediterranean Applicants: Turkey, Cyprus and Malta", *Bank of Valletta Review*, No. 7, Spring 1993.

Reinton, P. O. (1967), "International Structure and International Integration: the Case of Latin America", *Journal of Peace Research*, Vol. 4.

Riedel et al. (1994), "What Is Our Long Term Vision [in the Middle East]?", *Middle East Policy*, Vol. 3, No. 3.

Rostow, W. (1990), "The Coming Age of Regionalism", *Encounter*, Vol. 74, No. 5, June 1990.

Selim, M. El-Sayed (1994), "Mediterraneanism: A New Dimension in Egypt's Foreign Policy", Paper prepared for presentation at the *American Political Science Association*, New York.

Sneider, R. L. and Borthwick, M. (1983), "Institutions for Pacific Regional Cooperation", *Asian Survey*, Vol. 23, No. 12.

Snyder, J. C. (1993), "Proliferation Threats to Security in NATO's Southern Region", *Mediterranean Quarterly*, Vol. 4, No. 1, Winter 1993.

Spencer, C. (1994), "Algeria in Crisis", *Survival*, Vol. 36, No. 2, Summer 1994.

Swan, R. (1992), "France the Scapegoat", *Middle East International*, 24 January 1992.

Thompson, William R. (1973), "The Regional Subsystem", *International Studies Quarterly*, Vol. 17, No. 1, March 1973.

Time International, 14 June 1993.

Time International, 9 January 1995.

Tsoukalis, L. (1977), "The EEC and the Mediterranean: Is Global Policy A Misnomer?", *International Affairs*, Vol. 53, No. 3.

Vayrynen, R. (1979), "Economic and Military Position of the Regional Power Centres", *Journal of Peace Research*, Vol. 16, No. 4.

Vayrynen, R. (1984), "Regional Conflict Formations: An Intractable Problem of International Relations", *Journal of Peace Research*, Vol. 21, No. 4.

Vernon, R. (1994), "Research on Transnational Corporations: Shedding Old Paradigms", *Transnational Corporations*, UN Publications, New York, Vol. 3, No. 1, February 1994.

Waever, O. (1995), "Europe's Three Empires, A Watsonian interpretation of post-wall European security", paper delivered at the 36th Annual Convention of the International Studies Association, Chicago, 21-25 February 1995.

Wallerstein, I. (1993), "The World-System After the Cold War", *Journal of Peace Research*, Vol. 30, No. 1.

Weinland, R. G. (1979), "Superpower Naval Diplomacy in the October 1973 Arab-Israeli War: A Case Study", *The Washington Papers*, Vol. VI, No. 61, Beverley Hills, Sage Publications.

Yahia, H. B. (1993), "Security and Stability in the Mediterranean: Regional and International Changes", *Mediterranean Quarterly*, Vol. 4, No. 1, Winter 1993.

Young, O. R. (1968), "A Systematic Approach to International Politics", *Princeton University Centre of International Studies Research Monograph 33*.

Young, O. R. (1969), "Professor Russett: Industrious Tailor To A Naked Emperor", *World Politics*, Vol. XXI, No. 3, April 1969.

Zartman, I. W. (1967), "Africa as a Subordinate State System in International Relations", *International Organization*, 21, 3, Summer 1967.

International Organizations

Commission of the European Communities (1984), "On The Implementation of a Mediterranean Policy for the Enlarged Community", 28 March 1984.

Commission of the European Communities (1989), "Commission Opinion on Turkey's Request for Accession to the Community", SEC (89) 2290 final/2, Brussels.

Commission of the European Communities (1991), "The European Community and Mediterranean Countries".

Commission of the European Communities (1992), "The Future Relations Between the Community and the Maghreb", SEC (02) 401 final, 30 April 1992.

Commission of the European Union (1994), "Strengthening the Mediterranean Policy of the European Union: establishing a Euro-Mediterranean partnership", COM (94) 427 final, 19 October 1994.

Commission of the European Union (1995), "Barcelona Declaration", 28 November 1995, Final Version.

Conference on Security and Cooperation in Europe (1992), "Helsinki II Declaration", July 1992, Geneva.

Conference on Security and Cooperation in Europe, Doc/3-C/Dec. 1992.

Conference on Security and Cooperation in Europe (1993), "Mediterranean Seminar", Chairman's Summary, May 1993.

Conference on Security and Cooperation in Europe (1994), Review Conference statement made by Maltese Ambassador Maurice Abela, Maltese Ministry of Foreign Affairs, Press Release No. 472/94.

European Parliament Sessions Document (1993), "Report of the Committee on Development and Cooperation on Relations Between the European Community and the Maghreb", Ceretti, M. L. C. (1993) (SEC(92) 0401 final), 12 May 1993, Doc-EN\RR\227\227339.

Eurostat, "European Union Trade with the Mediterranean Countries", 10 November 1995.

Law, D. (1993), "A New Approach For The Mediterranean", North Atlantic Assembly Rose-Roth Seminar On Security and Cooperation in the Mediterranean Region, Granada, 2 February 1993.

Martinez Report (1991), "European Security and Threats Outside Europe - the Organisation of Peace and Security in the Mediterranean Region and the Middle East", *Assembly of the Western European Union Report*, Doc. 1271, May 1991.

Roseta Report (1993), "Security in the Mediterranean", Assembly of the Western European Union, Doc 1371, 24 May 1993.

WEU Report, 15 June 1994.

World Bank Report, "Claiming the Future Choosing Prosperity in the Middle East and North Africa", October 1995.

Newspapers

Financial Times, "Special Survey on Tunisia", 14 June 1993.

Financial Times, "Mediterranean Eurocorps Under Discussion", 27-28 December 1993.

Financial Times, "Hopes Smothered in Shifting Sands", 10 June 1994.

Financial Times, "Domestic Politics sets France apart on Algeria policy", 15 July 1994.

Financial Times, "Marginalisation of Mediterranean Countries", 15 November 1994.

Financial Times, "Israel and the EU", 17 December 1994.

Financial Times, "EU close to deal on entry by Cyprus", 2 February 1995.

Financial Times, "'Separation' mars Middle East Integration", 9 February 1995.

Financial Times, "Koor's Mr Turnaround builds bridges in the Middle East", 13 February 1995.

Financial Times, "Bit-part players take centre stage", 13 February 1995.

Financial Times, "Ministers hard pressed to keep up on defence front", 2 June 1995.

Financial Times, "Donors are in two minds about need for peace deal", 13 July 1995.

International Herald Tribune, "EU Turns Sights South Across Mediterranean", 21 October 1994.

International Herald Tribune, "UN Chief Chides Security Council on Military Missions", 6 January 1995.

International Herald Tribune, "Arab States Rule Out Regional Bank", 14 February 1995.

International Herald Tribune, "Identity Jolt for Egypt: Its Still Special, but Not as Special as Before", 16 February 1995.

New York Times, "Israelis and Arabs Talking Business", 1 November 1994.

New York Times, "UN Security Role", 6 January 1995.

Times (London), 19 May 1959.

Times (London), 10 May 1995.